Science of Weather and Environment

Chapter 1

Composition of Atmosphere

The most primitive man—the wildest savage—recognizes a wind when out in it, but it took a Greek philosopher to tell us what it is: air in motion. What motion is we know, at least well enough for practical purposes, but what is air? What is that invisible and odorless something we breathe and therefore call atmosphere, the thing that affords us all our weather perceptions and whose states and conditions we have learned to measure and even to foretell, and what was its origin?

ORIGIN OF THE ATMOSPHERE

To start as nearly as possible at the beginning, how and when did the earth ever get an atmosphere—its gaseous envelope? Well, there are two, and so far as we know, only two basic substances, the electron, a certain extremely minute quantity of negative electricity, plus something else, maybe; and the proton, an equally small quantity of positive electricity, plus something else, also maybe. The neutron or chargeless mass is omitted as so little is known about it. Of the origin of these entities no one has the slightest idea.

They can exist separately, in which case they are electrically very active; or variously grouped together in equal numbers, with increase of inertia but almost total loss of electric force, at least on things external. Every such group that is stable, or even measurably durable, is a chemical element. There are no other elements, and this number is limited. Furthermore, though occurring in unequal quantities, all these elements appear to be distributed throughout the universe. Most of those known on the earth have been found in the sun, for instance, and we believe the others are there, too, even if in such relatively small amounts as to be difficult of detection.

Hence, when that other star, according to one cosmic theory, some three billion four hundred million years ago, passed so near (within a thousand million miles, perhaps) our sun as to drag off from it by tidal action the masses that coalesced into the planets, there were present, and came off together, all the possible elements. That is, the primordial material of the earth, as it was pulled off from, or out of, the sun consisted of all the elements that now make it up, so that at the very beginning of the independent existence of the earth it had, if not an actual atmosphere, at least the makings of one.

It is believed, by those who hold to this theory that as the earth mass drew together in a molten sphere the heavier and more refractory substances formed mainly the inner core, while the lighter and more volatile elements and compounds formed the rocky shell and the gaseous envelope. In this way, they claim, the earth soon had oceans and an atmosphere of some kind.

However, even if all the gases, hydrogen, oxygen, nitrogen, and others, that can combine with various elements and compounds and form solids had then so combined, leaving neither water nor air, it is certain that before long the earth would have begun accumulating both. It is doing so now through every volcano, every fumarole, and every bubbling spring, and must always have done so since before even the first crust began to form.

But it seems unlikely that our present atmosphere actually did come entirely from the molten interior, whether by way of volcanic activity or otherwise, because volcanoes do not, so far as we know, give off free oxygen—volcanic gases captured before there has been any chance for admixture with the air show no trace of it. Furthermore this element could not exist uncombined in the presence of hydrogen and sulphur at high temperatures, both of which are abundant in volcanic vapors. But we have oxygen; where did it come from? It is known that highly developed green plants extract, under the stimulus of light, a great deal of oxygen from carbon dioxide, a gas abundantly emitted by volcanos. But most primitive plants do not, they consume it; hence this promising source of free oxygen appears, on close examination, to be quite uncertain.

However, some forms of lower life thrive in the absence of free oxygen and yet, in the presence of light, evolve it from certain of its compounds, especially water. Again, lightning, which must have occurred from the beginning, frees a little oxygen from water; and so also may ultra-violet light. There have been, then, continuously active means of obtaining free oxygen from its compounds since the beginning of the world. Nevertheless, it seems most likely that at the beginning there was more oxygen present than was necessary to use up the free hydrogen and to combine with the available surface materials —enough to do all this and to have a goodly amount left over as free oxygen of the atmosphere.

Presumably, too, there were present other primitive gases,

especially nitrogen, argon, carbon dioxide and water vapour. But whatever the primitive state of the atmosphere when the earth first formed, it may be regarded as practically certain that during the whole of the three billion four hundred million years since that time it has been continuously depleted by combination with many things in and of the crust, and also as continuously repleted by their decomposition. It is always changing, but, except in respect to water vapour, the change is so small in comparison with the whole that we are not ordinarily aware of it.

It has been argued that the could not have retained an atmosphere when molten, or even when dull red. But this is true of only the lightest two gases, hydrogen and helium, and of them only if no other gases were present in large amounts. A deep atmosphere of water vapour, for instance, would catch any light gas that might leave the earth beneath with an escaping velocity.

It could escape only if the outer portions of the atmosphere also were quite hot. At any rate, as soon as a crust, however thin, formed over the earth the supply of heat from beneath was so reduced (the crust being a good insulator) that the upper air necessarily became cool enough to retain the lightest gases. Also water must soon have begun gathering on the surface. Even if, up to this stage, all helium and free hydrogen had been driven wholly away from the earth, a condition that seems unlikely, there has been since then, and still is, abundant opportunity to accumulate both of them—hydrogen from volcanos, and helium from radioactive materials everywhere.

Presumably, therefore, the atmosphere is primitive in part —pulled off from the sun with all the other elements—and in part, at least, certainly regenerated inasmuch as every volcano is an active air factory.

Apparent Simplicity of Air

Presumably every one usually thinks of air as being just air, a homogeneous and single thing. Many of us always think of it that way, as the ancients did, when we think of it at all. Indeed so far as its behaviour and most of its physical properties are concerned it shows no obvious complexity. It is just air in motion that is responsible for a thousand familiar things from the stir of a leaf to the wreck of a house; and just air that floats the balloon and sustains the aeroplane. In all these matters it usually is quite satisfactory to regard the air as the single substance it ordinarily seems to be.

Evidence of Complexity

This apparent oneness of air does not, however, extend to all physical processes. When we try to liquefy it, for instance, evidences of its complexity soon become amazingly conspicuous. If untreated air is forced through the cooling coils they quickly become choked with ice; and they still clog up when even the driest air is used, if nothing but the water vapour has been removed—this time with solid carbon dioxide. Then, too, the liquid air itself shows abundant evidences that it is a mixture and not a simple substance like water. What then are the known constituents of the atmosphere, and how and when were they discovered?

Water Vapour

The earliest considerations of the composition of the air that have come down to us are those of the Greeks. In their speculative philosophy on the composition of objects, they considered air to be one of the four elements (fire, earth and water being the other three) that, singly or variously combined, make up all substances. From this it might seem that these Greek philosophers regarded the atmosphere as strictly a single thing—the "element" air. Yet it appears that by "air" they meant anything gaseous, and not necessarily the atmosphere. At any rate, Aristotle 250 years B.C., says very distinctly, in his work on meteorology, that cloud and rain are caused by condensation from the atmosphere of water vapour that had gotten there by the evaporation of water at the surface of the earth. He thus makes it very clear that the air consists of at least two things, and that water vapour is one of them.

Water vapour, then, was the first constituent of the atmosphere to be explicitly recognized. Aristotle mentions it, but it is not certain that this discovery was original with him. However, his is the earliest record we have of it, and for that reason, there being no evidence to the contrary, we regard him as one of the discoverers—the earliest one—of the constituents of the air.

DELAY OF FURTHER DISCOVERIES.

For more than 22 centuries, therefore, and perhaps for much longer, it has been known that the air we breathe consists of at least two things, water vapour and whatever is left after the water is removed. And for more than two thousand years after the days of Aristotle this is all that was known about its composition.

Indeed it was practically impossible to push our knowledge of the atmosphere any farther without something of the facilities and methods of the modern laboratory, nor before there had been acquired—very slowly and tediously it was—a fair concept of chemical elements and pure substances. Not until the beginning, then, of the 18th century was it reasonably possible for any constituent of the air to be discovered in addition to water vapour, nor indeed was any discovered until long after that. The chief obstacles that prevented such discovery for more than a hundred years after enough advance for that purpose had been made in laboratory technique, for that had been adequate from the beginning of the 17th century, were:

- The fixed idea that all gases are alike, all just air, and that any differences between various samples are due only to greater or less modifications of one and the same thing.
- The completely misleading and faulty concept that flame or combustion is the escape of something, phlogiston they called it, from within the burning object.
- The failure to recognize that change of weight incident to strong heating, during combustion, or under any other circumstances, was a matter of importance or had any scientific significance whatever.

A century before any constituent of the atmosphere, except of course water vapour, was recognized and collected in an approximately pure state, and while the faulty notions just listed were still prevalent, two people, working entirely independently, came near to finding one or more of its elements. The first of these was Robert Boyle (1627-1691), a wealthy bachelor, chemist and theologian; discoverer of the fact, known as "Boyle's Law," that doubling the pressure on a gas reduces its volume by one-half, of course for the same temperature. It seems very probable that Boyle would have discovered some of the constituents of the air if he had carried to completion certain experiments that he definitely listed. But there is nothing in his voluminous writings to show that he ever got them beyond the paper stage, despite the fact that, so long as his health permitted, he was a persistent worker.

The second near, even nearer, discoverer of certain of the atmospheric constituents was John Mayow (1643-1679), a graduate in law at Oxford, who turned to medicine and became noted as a physician, a chemist and a physiologist. After many and well-devised experiments Mayow concluded that the air consists of at least two portions, one that supports combustion and sustains life, and another part that does neither. The former he called "fire-air," because it keeps a flame going. He also said that it consists of "nitro-aerial particles," that is, particles in a gaseous form of the kind that makes a mixture of niter (saltpeter) and charcoal, or other combustible, burn, when lighted, in the absence of air—even under water.

All this is true enough, but his proof that the air consists in part of a special constituent, different from all the rest, was not complete. He did not collect "fire-air," the gas we now call oxygen, in a practically pure form and show that it is identical with the "fire-air" of the atmosphere. However, he recognized the incompleteness of some of his arguments, and it seems likely that if he had lived a few years longer his proofs would have been perfected, and our knowledge of the composition of the atmosphere set forward almost a hundred years. But this near attainment to the goal, and also even the direct route to it, appear to have been lost sight of for nearly a century. Indeed "fire-air," that came so near to being the first constituent of the atmosphere to be discovered, except, of course, water-vapour, and whose properties make it the most conspicuous, turned out to be the very last of all the major ones, save only argon.

Carbon Dioxide

The first of the permanent gases of the air of nearly constant quantity to be clearly discovered was not, as would seem most likely, either oxygen or nitrogen that together make up nearly the whole, but carbon dioxide, that is present as scarcely more than a trace—three parts in ten thousand. At that time alkalies were given to persons suffering with urinal calculi, and certain physicians recommended lime water for the same purpose.

Which the view of finding something still better Black undertook investigations with *magnesia alba*, a form of carbonate, as we now know, or, more exactly, basic carbonate, of magnesium. On strongly heating this substance a gas is given off, and Black turned his attention particularly to that gas, or air, as all gases were then called. He tried calcining, or burning to a powder, various substances, such as limestone, that we now know to be carbonates, and studying the gas thus obtained. In the end he found that this gas is much heavier than ordinary air, that it will not support life or combustion, that it will recombine with the calx, or powder, produced by the strong heating of the original substance, and that heat will again expel it as before. This gas seems to be fixed, or somehow fastened, in the objects from which it may be obtained.

Nitrogen

The composition of the air in a closed vessel after it no longer would support combustion or life, and to determine the cause of its unwholesomeness. After burning charcoal, phosphorus, or other combustible, in a closed volume of air, as long as possible, he removed the "fixed air" (carbon dioxide), if any had been formed, by means of lime, or an alkali, all in accordance with the previous investigations of Black, and then examined the remaining gas. He showed that this residue is not ordinary air because it supports neither life nor combustion, and that it is not fixed air for the alkalies do not absorb it. He called this residue "mephitic air," because it does not support life. We now know and say that this residue had been obtained by burning out the oxygen of the confined air and then absorbing the carbon dioxide thus (in most cases) produced.

And we know, too, that this residue, this "mephitic air," was nearly pure nitrogen. But at that time oxygen had not yet been discovered, and of course Rutherford could not talk in terms of things and chemical reactions then unknown. He did know, however, that it was obtained by combustion in an inclosed or limited volume of ordinary air, and therefore concluded, after the philosophy of his day, that it was atmospheric air combined with, or modified by the addition of, phlogiston—a mysterious fire substance whose escape from an object commonly is manifested by flame. Nevertheless, and no matter what his ideas as to its nature, Rutherford did obtain reasonably pure nitrogen and did record some of its properties, and therefore may be regarded as the discoverer of this constituent of the air, the most abundant of all.

Oxygen

Almost immediately after the discovery of nitrogen, the other major constituent of the atmosphere, oxygen, was independently found by Joseph Priestly. Priestley was a preacher by occupation and a chemist for recreation. At one time, while waiting for the building of the parsonage to be finished, he had the good fortune to live next door to a brewery—good fortune, because he was induced thereby to take up the chemistry of gases, his chief avocation for many years, on which he published several volumes, and which made his name famous. In the course of this work he obtained a red powder by heating mercury in the presence of air, and then on more strongly heating this powder with a burning lens he got a gas which supported combustion much better than ordinary air.

He had gotten oxygen, as we now know, by first burning mercury to an oxide and then decomposing that oxide by raising it to a high temperature. This element of the atmosphere he called dephlogisticated air. Combustible things burned in it readily; their phlogiston, or fire principle, rushed into it as air into a vacuum. It was air, that is, a gas, but air deprived of phlogiston. But regardless of what he called it

Priestley had discovered a new constituent of the atmosphere, the one we now call oxygen, the one that sustains life and supports combustion. On the other hand, Scheele was a professional chemist, so completely absorbed in his subject that he had little time for anything else, whether occupation or diversion.

His incentive to study the air was his wish to solve the riddle of fire, a thing essential to so many chemical processes. He found substances, such as phosphorus, which on burning reduced the volume of the confined air (at the same temperature and pres sure) in which they were burned by about one-fifth. In all cases the remaining air would not support combustion. He also found many ways of getting a gas that would support combustion far better than ordinary air, that was heavier than the air left after combustion, and that made burnt air indistinguishable from ordinary air when mixed with it in the proportion of one volume to four, or thereabouts. The residual or burnt air, the nitrogen, essentially, as we now name it, he called "vitiated air." The other portion, the part that supports combustion, he called "fire-air," just as Mayow had called it a century before.

Scheele, like Priestley and every other chemist of his day, except the immortal and tragic Lavoisier, interpreted fire phenomena in terms of phlogiston. He therefore considered "fire-air," which we call oxygen, to be a combination of phlogiston and a subtle acid substance. But whatever his opinions may have been as to its possible composition, he did find this constituent of the atmosphere, and that too even before it was found by Priestley.

Another student of the atmosphere that must be mentioned in connection with the discoveries of its composition was Henry Cavendish. He may not generally be credited with the discovery of any one of these constituents, and yet he appears to have found nitrogen about the same time that Rutherford did, if not earlier, by the simple process of passing the same confined air back and forth over red hot charcoal and then removing the fixed air (carbon dioxide) with an alkali. He did not publish this—he appeared always to be indifferent about publishing his investigations—but described it in a letter to Priestley in 1772, the year Rutherford published his discovery of mephitic air.

Lord Rayleigh had been making careful measurements of the densities of different gases, and thus came upon the fact that the nitrogen of the atmosphere, or rather that residual then regarded as pure nitrogen, was a little denser than the nitrogen obtained from any one of several chemical compounds. This was capable of three or four different interpretations, one of which, then regarded as the least likely, was the presence in the atmosphere of an unknown gas denser than nitrogen. At that stage Sir William Ramsay joined Lord Rayleigh in a common attack on this problem. Each tried removing from the air the then known constituents, using entirely different methods for getting rid of the nitrogen, and each found conclusive evidence of the presence in the atmosphere of an unknown gas to the extent of about one per cent. It proved to be chemically inert, and for that reason was called argon.

Other Inert Gases

During the years 1895-1898, that is, immediately after the discovery of argon, four additional inert gases, helium, neon, krypton and xenon, were found by Sir William Ramsay and his assistant, Mr. M. W. Travers, to be constituents of the atmosphere. Helium was found as that portion of the atmosphere that is still gaseous while the rest has been chilled to liquids or solids; and the others, neon, krypton and xenon by fractional distillations of large quantities of liquid air and liquid argon, impure, of course.

HYDROGEN

Lord Rayleigh and others also found free hydrogen always present in the air, but in amounts that varied from, roughly, one part in 1,000,000 to one 5,000. Perhaps this gas can not be regarded as a permanent component of the atmosphere in the same sense that oxygen and nitrogen are, but rather as an accidental and variable impurity. At any rate it is irregularly added to the air, at least by volcanoes, and more or less irregularly removed from it.

Traces and Impurities

Varying traces of ammonia, nitric and nitrous acids and their compounds, sulphuric and sulphurous acids and their compounds, oxides of nitrogen, hydrogen dioxide, and ozone are among the innumerable things always in the air if not of it. Minute particles of sea salt from evaporating spray, fine earth dust caught up by winds, pollen of every description and spores of many kinds are among the coarser and ever present pollutions of the atmosphere from the earth beneath. Shooting stars furnish a continuous, though invisible, shower of dust the world over from outer space. Also radioactive products of radium, thorium and other elements are pouring into the atmosphere continuously and contributing to the maintenance of its electrical state. This state means, in part, around 20,000 electrified particles, or ions, per cubic inch in the lower air. The number of ions 60 miles or so above the surface is far greater, at least 1,600,000 per cubic inch, as we infer from the phenomena of radio communication. In short, while the things of the atmosphere are but few the things in it are innumerable. All these latter are relatively very small in amount, but some of them are exceedingly important. The salt particles and some others are essential to condensation and rainfall; the ammonia and other nitrogen compounds, brought down by precipitation, add much to the fertility of the soil; the ions of the high atmosphere make distant radio communication possible; and ozone, also in the upper air, shields us from that portion of the ultra violet radiation that would destroy our eyesight, as we are now constituted, and otherwise do us irreparable harm.

PERCENTAGE COMPOSITION OF PURE DRY AIR

If the atmosphere is freed from all its numerous impurities and also freed from water vapour, there will remain pure dry air whose percentage composition at the surface of the earth is, very closely:

Nitrogen	78.03
Oxygen	20.99
Argon	0.9323
Carbon dioxide	0.03
Hydrogen	0.01
Neon	0.0018
Helium	0.0005
Krypton	0.0001
Xenon	0.000009

The amount of water vapour in the atmosphere, or better, perhaps, the amount of the water vapour constituent of the atmosphere, varies from scarcely more than a trace, at extremely low temperatures, to at least 5 per cent by volume on the hottest and most humid days. At and above the height of 5 miles, say, the amount of water vapour always is small, even when saturation obtains, owning to the very low temperatures at these levels. That is, water vapour is confined almost wholly to the lower atmosphere. Its average value, the world over, is such that if it were all condensed it would be the equivalent of a layer of water about one inch deep over the entire earth.

FORMS OF OXYGEN AND THEIR DISTRIBUTION

Oxygen also is peculiar in its distribution, and it occurs in three different forms. All but about one part in 400,000 is ordinary oxygen, or, in the language of the chemist, diatomic oxygen. Triatomic oxygen, or ozone, occurs chiefly beyond the highest clouds, its greatest density being, apparently, at the levels of 20 to 30 miles. It occurs in the lower air only as a trace. In the highest air, 60 miles and more above the surface, the oxygen appears to be at least partly monatomic, according to the spectrum of the aurora.

Importance of Ozone

Every one knows that life would be impossible if there were no ordinary or diatomic oxygen in the atmosphere, and that without it nearly if not quite all vegetable life also would be impossible. But it is not so well known, though equally true, that animals, including the human species, could not exist as now constituted if the air did not contain a small amount of triatomic oxygen or ozone. And yet, paradoxical as it may if seem, ozone were just a little more effective in its goodness, again life, as now constituted, could not last.

These surprising facts come about in this way: The radiation from the sun includes not only every colour, that is, the whole of the visible spectrum, but also extends indefinitely beyond the red into the long-wave-length invisible region, and likewise, in the opposite direction, well beyond the limit of the violet. But in this ultra violet portion of the spectrum the radiation from the sun ceases far short of the limit to which that from an electric arc, for instance, can be followed; and ceases, not because no radiation beyond that limit is given out by the sun, but because it is absorbed by the ozone in the upper atmosphere.

Now, much of that particular radiation which ozone absorbs is destructive to the eye, and when intense probably injurious to other tissues as well. On the other hand, it also absorbs a large part of the radiation that is effective in preventing rickets, but, and this is of the utmost importance, it does not absorb quite all of this antirachitic portion.

Enough is let through to keep us in proper health. This particular limiting, then, of the solar radiation that reaches the earth is amazingly located. If it was a little farther out in the ultra violet, eyes, as now constituted, could not have developed; and if a little nearer the visible, again, animals, as they now are, could not have come into being. Of course ozone, in moderate amount, presumably was in the air long before there were animals of any kind to be affected by its action on solar radiation.

This radiation, therefore, was not adapted to them, but they developed in adaptation to it; nevertheless the fit is so close and of such a surprising nature as to give us decided pause for thought.

Origin of Ozone

Ozone is produced by the action of extreme ultra violet light on oxygen, and therefore at great heights; a little where the oxygen is very rare, then more and more with decrease of height and increase of the oxygen supply until by absorption the ozonizing rays are considerably enfeebled. As the density of the oxygen increases the intensity of the effective radiation decreases, hence the rate of production of ozone by this process must be very slow in the outermost air, increase to a maximum with decrease of height and then rapidly fall off at still lower and lower levels. The total range of ozone, top to bottom, probably is not less than 100 miles, with its maximum concentration around 25 miles, perhaps, above the surface, and yet all told, and despite its great importance, it is the equivalent of a layer of this gas only about one-tenth of an inch thick at atmospheric pressure and room temperature. Ozone is produced also by electric discharges through oxygen. Lightning certainly produces some ozone, and it may be that the auroras form it too, though the great heights at which they occur, 60 miles and more above the earth, seem to render this conclusion doubtful.

Distribution of Ozone

As already stated, we know that at most there is only a trace of ozone in the atmosphere up to the level of the highest clouds, and that it exists to an appreciable extent at considerably greater heights. But this is not all. The total amount of ozone vertically over one to the limit of the air appears to increase, in general, with increase of latitude; to be greater during winter and spring than summer and fall; and to be greater in winds from high latitudes than in those of the opposite direction. This complicates the problem of the origin of ozone. If it is produced by ultra violet radiation alone why should it not be most abundant in tropical regions, and elsewhere in the summer time? If we surmise that it is produced largely by auroral discharges, how, we ask, can these discharges, 60 miles up and more, reach a sufficient supply of oxygen? And if there is enough oxygen at these great heights how can the ozone subsequently become concentrated at the intermediate levels? These are some of the questions to be answered by future observations and studies.

UPPER AND LOWER ATMOSPHERE

The composition of the lower atmosphere up to at least 6 or 7 miles in temperate regions, and 8 or 9 within the Tropics, is well known, except in respect to condensation nuclei and certain impurities. Throughout this region too the percentages of the several gases, except water vapour, are practically constant owing to their continual mixing incident to convection and turbulence.

Somewhere in the upper air, however, the percentages of the lighter gases must increase and those of the heavier decrease under the action of gravity since at these levels vertical convection is very feeble. But as the upper air grows thinner the more and more rapidly does our knowledge of it become less. We know that its outermost portion is the very extensive region of the aurora, hundreds of miles thick; that near the base of the auroral region there is enough ionization to make wireless communication around the world entirely practicable; and that far below this Kennelly-Heaviside layer, in turn, and yet well within the upper air occurs most of the ozone, the triatomic oxygen that indirectly is so vital to all terrestrial life. We seldom give any thought to this upper air, but it is so important that we really must know more about it.

Mass of the Atmosphere and of its Constituents

From the known percentages of the several constituents of dry air, given above, their molecular weights, and various other pertinent facts such as the amount of water vapour present, height of the barometer, volume of land above sea level, and distribution of temperature with height, it is easy to compute the approximate mass of the atmosphere as a whole and of each of its several gases. The results are given in the following table, in which the factor 108 means: Add eight ciphers, or multiply by 100,000,000.

Since the values in this table were determined more or less independently it could not be expected that the percentages found of the constituents would add up to exactly 100, nor that the sum of the computed masses of the several parts would precisely equal the mass of the whole. These deviations, however, are very small—probably within the present limits of experimental errors.

The numbers here given that express the masses of the atmosphere and its several constituents are useful as quantitative values and for exact comparisons, but so great, even though in terms of tons, that we can form no distinct conceptions of them—they are just awfully big! A clearer idea may be gotten from the fact that the total mass of the atmosphere is the equivalent, roughly, of that of a block of granite a thousand miles long, a thousand miles broad and half mile thick; while the least abundant of the constituent, xenon, if loaded on cars, 19 tons to the car, would freight a train reaching 40 times around the earth along a great circle, and which, traveling 20 miles all hour, would be 6 years in passing any point on the road.

Table. Mass of the Atmosphere and of its Constituents in tons (2000 pounds)

Substance	Volume per cent dry air, at surface	Total mass
Total atmosphere		56,328,000 × 108 tons
Dry air	100.00	56,181,850 " " "
Nitrogen	78.03	42,684,725 " " "
Oxygen	20.99	12,782,647 " " "
Argon	0.9323	682,125 " " "
Water vapor		146,150 " " "
Carbon dioxide	0.03	23,874 " " "
Hydrogen	0.01	1,423 " " "
Neon	0.0018	759 " " "
Krypton	0.0001	141 " " "
Helium	0.0005	88 " " "
Ozone	0.00006	33 " " "
Xenon	0.000009	19 " " "

Chapter 2

Troposphere and Stratosphere

After discussing the composition of the atmosphere it would seem appropriate next to consider its structure. "Structure?" we ask. "Wouldn't a chapter on the structure of the atmosphere be much like the famous chapter on snakes in Horrebow *Natural History of Iceland*—'There aren't any'?" Yes, it would seem so, we must admit, to any casual observer, as it did to every one until only a short while ago, comparatively. Now, however, we already know of various ways in which the air is streaked, laminated and otherwise divided into distinct portions, and presumably there still are many more parts, according to this or that basis of division, that yet will be found and studied.

At about the close of the last century, soundings of the atmosphere were begun with the aid of small balloons carrying light devices that automatically registered the temperature and pressure throughout both the ascent and the descent—a kind of exploration that soon was taken up at various places. From these records, in turn, the heights corresponding to given points on the traces were readily computed. In this way we gradually have come to know the average temperature and pressure of the air at every height from the surface of the earth up to at least 12 to 15 miles, under different weather conditions, for all the seasons, and in many parts of the world. Pretty soon a means of registering the humidity was further added to the apparatus carried, so that we now have a fair knowledge also of the average vertical distribution of water vapour that corresponds to each particular type of weather. Obviously, too, these sounding balloons, as they are called, afford some knowledge of the direction and velocity of the wind at various levels for the particular time and place at which a flight is observed.

From the data thus obtained several interesting generalizations soon became evident. The most conspicuous of all, and for a long while the most doubted, because it was not understood, was the fact that at 6 or 7 miles above sea level, in middle latitudes, and generally 8 to 10 in tropical regions, the temperature no longer decreases with increase of height, but remains substantially constant. How far this equal temperature, or isothermal condition, extends is, of course, unknown, but it does go at least to the greatest altitudes yet attained, that is, 15 to 20 miles.

It may extend to the limit of the atmosphere of appreciable density, or it may not. We have no direct and positive evidence of either alternative. This much we know. From the surface up to a considerable distance above sea level, generally 6 to 10 miles, depending mainly on the latitude, the temperature decreases at the average rate of about 1° Fahrenheit in 300 feet; then almost abruptly, as a rule, practically ceases to change with further ascent. Clearly, therefore, this unsuspected and striking phenomenon divides the atmosphere into two great parts; a lower portion in which temperature rapidly decreases with increase of height, and an upper in which the temperature is nearly independent of height.

In the lower, convection, or ascent and descent of the air, can occur under certain conditions, because although ascending air cools by expansion (the pressure on it being decreased by the weight of the air left below) its ascent brings it into air that also is colder. Sometimes the ascending air, especially when saturated with water vapour, cools less rapidly with increase of height than does the air through which it is passing, in which case convection is certain and often vigorous. Similarly, a local mass of air, particularly when cloud-laden, may warm less rapidly on sinking, as such masses do just after sun down, than the air it is falling through, whereupon convection again is inevitable. In short, this lower, cooling-with-altitude portion of the atmosphere is a region of convections and overturnings—not at all times, but under humidity conditions that will be explained later. It therefore has been called the troposphere, the sphere, or spherical shell, in which turning (going up and then going down), or convection, occurs.

In the region above the troposphere, where the temperature is constant with height, marked or vigorous convection does not and can not occur. It can not occur here because rising air rapidly cools with ascent, owing to expansion incident to decrease of pressure on it by the weight of the air passed through. Such cooling would keep the rising air all the time colder and therefore heavier, volume for volume, than the surrounding air and thereby quickly reduce its motion first to zero and then reverse it.

Formerly this was called the isothermal region, and to some extent it is still so called, owing to the fact that vertically, so far as explored, the temperature is substantially constant. Since convection here is impossible masses of air forced into this region would spread out in horizontal strata, and indeed there commonly are evidences of the existence of just such strata, though we seldom if ever are sure of their origin.

This filler structure suggested the other, and now the all but exclusive, name, "stratosphere," of this region. Thus the troposphere and stratosphere are quite distinct from each other. Clouds and every sort of precipitation, rain, snow, graupel, sleet and hail, and every kind of storm, are forever agitating the former, but never for a moment disturb the serenity or overcome the stability of the latter. They are sharply separated, the one from the other, along an approximately horizontal but invisible surface called the tropopause, and so called because that is where convection ceases, as above explained.

Perhaps it will be interesting to recall here that sometimes we use the expression, "the sphere," to mean the earth as a whole; that in our first approach to particulars we divide the sphere into the lithosphere, or solid portion of the earth, the hydrosphere, or water portion, and the atmosphere; and that the atmosphere in turn consists of two great concentric shells, the troposphere and the stratosphere. The lithosphere and the hydrosphere also have interesting structures about which many a fascinating tale has been told, but this is not the place to repeat them; ours is the story of the atmosphere and of it alone.

Trades (trade winds). When we examine the troposphere closely we find that it too has structure. Some of these parts are fleeting, but others are at least semi-permanent. One of the more nearly constant is the trade wind, or, better, the trade winds, as there are several winds properly so designated. Advantage has been taken of them, of course, in shipping, especially by sailing vessels, but this is not the source of their name.

It does not derive from any idea of commerce, but from the fact that their characteristic is persistent blowing along a particular trade, that is, tread, track or way. The trades are east winds (winds from easterly points) over the tropical oceans, or, more exactly, between the latitudes 30° N. and 30° S. There really are five such winds that are well defined, one over the north Atlantic Ocean between latitude 30° N., roughly, and the Atlantic doldrums, or region of calms near the equator, and another over the south Atlantic between latitude 30° S., also roughly, and the same doldrums. Two other trade winds are similarly located over the Pacific Ocean, and there is one over the Indian Ocean, south of the equator, between Australia and Madagascar.

These wind currents, the trades, are very shallow near their poleward boundaries or edges, but grow gradually deeper and deeper to a maximum of at least 4 or 5 miles as their equatorial borders along the doldrums are approached. They are well defined and important structures of the troposphere.

Antitrades

Immediately over the trade winds are the antitrades, or winds that blow away from the equator and at the same time turn more and more nearly eastward and come down lower and lower with increase of latitude until they merge with, and become a part of, the prevailing westerlies-the winds beyond latitude 30° N. or 30° S., as the case may be, that from the surface up to at least well into the stratosphere usually are from westerly points.

The antitrade (there is one over each trade wind) represents a part of the return to higher latitudes of its accompanying trade, which through most of its route approaches closer and closer to the equator. The rest of the trade, joined by a greater or less amount of tropical air, turns around the western end of a high pressure ridge, or ocean area of prevailing light and variable winds and clear skies. The frequent and deep southerly winds onto the United States from the Gulf of Mexico and the Atlantic Ocean are of one of those great joint trade and tropical currents.

Here then are two major features or elements in the structure of the troposphere, the trade and the antitrade, the east wind and the west wind, that all who pilot the argosies of the skies must know, and know how to use to proper advantage.

And the troposphere has many other parts. Two of the strangest and most important of these, strange because though apparently impossible companions they yet are always together, are the cold wind and the warm wind that peacefully flow beside each other—that, in technical terms, are each in dynamical equilibrium with the other. We know that a column of warm air will not stand up in equilibrium with an adjacent column of cold air, and we know that we know it by the way cold out-doors air pushes up and away the warm air in a heated chimney or smoke stack.

Hence we are unwilling at first to believe that a cold wind can blow alongside of a warm wind and not drive it away—that the twain can meet on equal terms and each hold its own.

But they can and do, because an object, such as a mass of air, whenever moving over the earth tends always, because of the earth's rotation, to turn to one side or the other (the right, going with the wind, in the northern hemisphere, the left in the southern) of the course it is on.

And the magnitude of this tendency (the force the moving object would exert on a frictionless vertical surface that would hold it to a fixed geographic direction) is proportional both to its mass and to its velocity. In the case, therefore, of adjacent stationary columns of warm and cold air only gravity is operative and the heavier (because colder) column underruns and buoys up the lighter.

This is why air rushes up heated chimneys and smoke stacks—why they draw, and the higher the better. In the case of the winds, however, the situation is quite different, for in proportion to their mass and their velocity they now exert (or would, against a suitable restraining wall) a horizontal or deflective force that also must be considered.

If, then, the winds were in the right direction, and had the proper positions with reference to each other, the colder from the east, say, and the warmer just south thereof and from the west, or, more generally, if they were passing each other counter clockwise (in the northern hemisphere, clockwise in the southern) they might have such velocities that the resulting deflective force would just balance the difference in pressure due to difference in density.

Ordinarily this condition of equilibrium implies a very gently sloping interface between the warm current and the cold, with the former overrunning the latter. Owing, however, to surface friction and various obstacles, such as islands, and mountains, to the free flow of the winds, and to a change, for one reason or another, in the relative amounts of air on the two sides, perfect equilibrium never is long maintained. It does not break down abruptly and completely, but gives way quite slowly, and in so doing often leads to the development of the general cyclonic storm of the middle and higher latitudes.

Cold Front

When the breakdown between the warm and the cold currents is well developed, with the mass of cold and, of course, polar air pushing its way equatorwards while the warmer air flows beside it poleward, we speak of the interface between them, or, rather, of the intersection of that interface with the surface of the earth, as a cold front. We also call this locus a windshift line because here the direction of the wind changes, as one system, the warm winds, departs and another, the cold, comes on.

It likewise is called a squall line because all along it the winds are turbulent and squally and, in summer, often accompanied by thunder storms. This cold front, then, is the boundary, along the surface, between distinct wind systems—between a system of relatively cold and dry winds under skies and from higher latitudes, and a system of comparatively warm and humid winds under clouded skies and from lower latitudes—two more of the great parts in the structure of the troposphere.

Warm Front

The warmer of the two passing winds just mentioned soon reaches, in the course of its poleward travel, the colder air of higher latitudes. But here, as elsewhere, two masses of air, having different temperatures can not be in equilibrium with each other at rest and standing side by side. They must be, and are, in motion, and in such manner as to retain for a time, and to a greater or less extent, their individuality. Here again the colder air is wedged under the warm which flows not along this wedge, but up its slope—slantingly as a rule, but up, nevertheless. The line along which this part of the interface between the cold and the warm winds cuts the surface is called the warm front.

Thus the cold and the warm sectors of the traveling cyclone maintain their independence as polar winds and tropical airs, respectively, until gradually separated from their source, after which they soon are brought into like conditions each with the air of its new location and merged with it. In this way polar air becomes tropical air, and tropical winds polar winds, back and forth ceaselessly and indefinitely; but always as entities in their travels—as elements in the structure of the stratosphere—just as rivers and lakes are entities in the ceaseless round of water from oceans to continents and continents to oceans.

Inversions

Normally the temperature of the lower atmosphere, or troposphere, decreases at every level with increase of height, but there are exceptions to this rule; and, besides, the rate of increase always is more or less irregular. A very common exception occurs next to the surface. Indeed, during still clear nights this exception is itself the rule, and practically without exception.

At such times, owing to the rapid net loss of heat by the surface through radiation, the adjacent air, by contact with the cold surface, becomes cooled to a distinctly lower temperature than the air at a slightly higher level. The height to which this cooling extends depends upon the amount of air movement and consequent turbulence or mixing. If this is slight the cooling is restricted to near the surface, but is all the more pronounced. If the movement of the air is appreciable the cooling is less pronounced but extends higher. Finally if there is considerable wind the mixing so distributes the loss of heat as to be small at any level, even at the surface. When, however, the cooling is marked and restricted to a shallow layer the temperature increases through this layer with increase of height instead of decreasing, as is the rule, from the ground up. This is an inversion of the usual temperature gradient, or a temperature inversion, or, for short, just an "inversion."

For a while the air next above a surface inversion layer can slowly blow along over it without rapidly mixing with the colder air or carrying it away thus providing another case of structure in the troposphere. Appreciable winds, however, do wear away such a layer rather rapidly. They also commonly tend to produce, and frequently do produce, a temperature inversion some distance, perhaps a thousand feet or more, above the ground.

The surface inversion, discussed above, occurs, as explained, on still clear nights and is due to loss of heat by radiation, and therefore might be called a radiation inversion. The inversion now under consideration may occcur any time the wind blows, because it is due to the turbulence or vigorous mixing of the lower air incident to surface friction and the interference to free flow by trees, house and other irregularities. Turbulence produces a temperature inversion —the turbulence inversion—in this way: Before the wind sets in, the temperature of the lower thousand feet or more of the air may and often does, decrease slowly with increase of height, perhaps only one degree Fahrenheit in 400 to 500 feet.

It even may increase with height, especially of early mornings, through the lowest levels, as just explained. In either case a complete mixing of the air from the surface up to the level of 1000 feet, says, brings all the air, not to a common temperature, as the stirring of water smooths out any thermal inequalities it may have had, but to such temperature that wherever, within that layer, a portion of its air may be taken, up, down or sidewise, it will, on arrival, have precisely the same temperature as the then surrounding air at the same level. That is, it will come to such temperature that nowhere will a moving portion of its either give heat to, or take heat from, the air with which it at any time is in contact.

Obviously, because if the agitated air is not yet in a state of temperature equilibrium with a moving portion of itself, all that need be done to make it so is to stir it up further and mix it more thoroughly.

Now an isolated mass of air that neither gains heat from nor loses heat to the atmosphere through which it passes, obviously must cool with ascent, owing to its loss of heat incident to its expansion against the decreasing pressure. Where the air is unsaturated, so that no fog or cloud is formed, and thoroughly stirred up, as above explained, the rate of this decrease, and therefore the rate of temperature decrease in and of the layer itself is, approximately, 1° F. per 190 feet, allowing for the average amount of humidity. This, as explained, is a much faster rate of decrease of temperature than usually exists in the lower air. Hence the mixing of the lower air by turbulence so redistributes its heat as to make the lower portion warmer than it was, or otherwise would be, and the upper portion colder. Immediately above the topmost portion of the turbulent layer the air is undisturbed by vertical convection and therefore distinctly warmer than the upper portion of the agitated stratum. The turbulence inversion (inversion caused by turbulence) is a sharp partition between two portions of the troposphere, one the lower, full of irregularities, the other smooth-flowing.

Owing to its comparatively low temperature the top of the turbulence layer often is covered with a broad but shallow cloud of the stratus type which, because located at an inversion level, is warmest over its upper surface. But whether clouded or clear this inversion level is difficult of passage by air from either side. If air should pass this level going up it would at once be surrounded by other air much lighter, because warmer, than itself and therefore it would drop back. Similarly, air passing it from above would be promptly pushed up again by the denser, because colder, atmosphere it was replacing. In short, this inversion stratum, though very thin and commonly invisible, is an impassable ceiling to rising air from below and an impenetrable floor to falling air from above.

Gradually, however, through heat conduction, thermal convection, and in other ways, the inversion is smoothed away, and interchange across this level thus made possible wherever the air has within and of itself an adequate supply of water or, really, steam power—wherever its water vapour is sufficient to give abundant condensation, as in a cumulus cloud, and thereby a quantity of heat sufficient to keep it all the way to great altitudes continuously warmer and lighter than the surrounding medium.

It is interesting that as the temperature changes almost abruptly, perhaps several degrees, with change of height at the level of this turbulence inversion, so also must the density of the air change abruptly. And, furthermore, if the pressure gradient or push that causes the winds, is practically the same, as it seems to be, on either side of, and close to, this level, then there also must be here a nearly sudden jump in the wind velocity, from slower in the under, denser air to faster in the lighter air immediately above. However, this change of velocity, seldom more than one per cent of the whole, always is too small to be of any practical importance.

OVERFLOW STRATA

As explained above, any layer of air that is thoroughly mixed up has a certain rate of decrease of temperature with increase of height, and such that the temperature of an isolated mass of like air rising or falling through it will change at the same rate. In such a layer vertical convection is as easy as horizontal gliding over a smooth surface. Left to itself, though, and if unclouded, its vertical temperature gradient or lapse rate (lapse, for short) gradually changes, owing largely to gain and loss of heat by radiation, until it becomes decidedly less, and the layer thereby impenetrable to dry or unsaturated air. Saturated air, however, may be buoyed up to considerable heights, as already explained, because the heat of condensation, or heat rendered sensible as a result of condensation, so reduces the lapse rate that the ascending air is warmer and therefore lighter than the air surrounding it.

The extent to which this rising air is warmer and lighter than the air through which it is passing depends, of course, on the amount of condensation. This in turn depends on the amount of water vapour present, and that depends on the temperature. Hence, in general, and starting from the same level, saturated warm air is pushed up to greater heights in the process of convection than is saturated cold air. In any case, though, the amount of condensation per given increase of height, and therefore the quantity of heat available for further convection, becomes less and less with gain of altitude, and finally, at one level or another, insufficient to induce further ascent.

At this level then, whatever it is, the rising air spreads out in a sheet or stratum that differs in humidity, temperature and lapse rate from the atmosphere of every other level, both higher and lower. In this way, that is, from convections, great and small, including the over- and underrunning associated with general or cyclonic storms, the troposphere is largely built up of overflow strata. They are not, of course, the same from day to day nor from place to place, but everywhere they are always more or less distinct and numerous. They often mark the levels of cloud layers of the sheet or stratus forms, of alto-cumuli and of the windrow or billow clouds, due to the waves caused by the flow of one stratum over another, much as water waves are induced by wind.

Occasionally, two adjacent strata differ from each other so radically that a balloon can float a long distance with the bag in the one and the basket in the other. Of course the identity of each particular stratum ultimately is lost through mixing with others above and below it, whether caused by the vigorous stirring incident to a general storm, or by virtue of the perpetual diffusion and prevalent turbulence over every interface. But so long as it does exist it may be pushed up bodily to greater heights by underrunning air, or depressed to lower levels by an overflow current, with, in either case, a change in the temperature gradient or lapse rate (except in the very unusual case when it initially is that of completely stirred-up air) and a corresponding change in its stability and resistance to penetration by convection in either direction, upward or downward.

If the layer is pushed down, without lateral contraction or expansion, by an overflow of air above it the pressure on it will be increased by the same amount throughout, but it will be compressed most, and thereby heated most, on the upper side where the initial pressure is least, and compressed and heated least on the under side where the initial pressure is greatest. This changes the temperature gradient in the stratum so depressed, except rarely, as above explained, and in such manner as to render the layer increas ingly difficult of penetration—a firmer floor and a more rigid ceiling. On the other hand, if the stratum is lifted to a higher level it is cooled most on top and least at the bottom and its effectiveness as a barrier to convection corresponding decreased.

It should be noted that in atmospheric convection, and the consequent production of air strata, water vapour plays a most important role. With increase of humidity, under constant temperature and pressure, the density of the air steadily decreases, just as it would with increase of temperature at constant pressure. This fact probably accounts for many small waterspouts starting at the surface—starting there because the lower air becomes relatively light through high humidity, analogous to the starting of dust whirls over a dry region due to decrease of density incident to increase of temperature.

There is, however, a fundamental difference between the surface waterspout and the dust whirl. The latter consists of dry air made light by increase of temperature, and can ascend (be pushed up) until it has lost a certain amount of its original heat and no further. Not so with saturated or highly humid air.

It, too, like the dry air, ascends because it is lighter than the adjacent air around about, but it maintains this relative lightness a much longer time through the latent heat rendered sensible by progressive condensation, and thereby reaches far higher levels. In respect to convection dry air and humid air are like unto two men in business, one with a working capital but no reserve assets; the other having, in addition to his ordinary current needs a much greater fund that may be drawn upon whenever required. The one, like dry air, may start well, but his power to expand soon is exhausted. The other, by drawing on his reserve, can take advantage of every opportunity and thus rise to a far higher level of success. Just as it takes money to rise high in the business world, so too it requires water vapour to make any considerable ascent in the atmosphere.

The konisphere (dust sphere) and its layers. Not all we breathe is air. With every breath we inhale a million microsticks and -stones and a host of other things that are no part of a pure atmosphere. "Where do they come from?" The heavens above and the earth beneath. Every wind that sweeps a desert catches up tons, and sometimes millions of tons, of pulverized rock to spread far and wide. Fragments of vegetable fibre litter the soil the world over and are wafted hither and yon as even the gentlest breeze may blow.

Pollen of conifers, ragweeds, and a thousand other trees and plants we must take into our lungs from spring to fall every day we breathe the open air. And our bronchial tubes need chimney sweeps (luckily provided by Nature) to get rid of their coatings of soot from kitchens, factories and forest fires. Even the ocean, through its evaporated spray, makes a salt mine of the air that we breathe. Then, too, lightning sprays nitrogen acids into the atmosphere, while soft coal and volcanic vents similarly add the sulphur acids—but all are too dilute really to bother us. Spores and microbes of many kinds we just have to inhale, for they are everywhere.

And as if all this were not enough the earth, every now and then, explodes at some great volcano and hurls tons upon tons of rock powder into the air where it drifts far away for weeks, months or years, according to its degree of fineness and initial height attained. Finally, in addition to all this dust of its own the world stirs up, the atmosphere to its outermost limits is filled with the ashes, so to speak, of daily millions of incinerated meteors, or shooting stars. That is how the earth got its konisphere (dust shell). If it had no atmosphere it would have no konisphere, but having an atmosphere it must also have a coexistent and coextensive konisphere.But this konisphere is not uniform; it has distinct layers that, like other phenomena, show structure in the atmosphere. The more pronounced of these layers are:

- The turbulence layer, that is, the layer of air next to the earth that, owing to surface friction, any appreciable wind fills with turbulence. Incident to this churning up of the air there also is a stirring up of the dust. At such times this is thedustiest portion of all the atmosphere, and it carries the largest particles. Its depth is, of course, that of the turbulence, and therefore may be anything from two or three hundred feet up to two or three thousand; while the amount of dust, as determined by the strength of the wind and condition of the surface, can vary from practically nothing at all, as over snow fields, to that of the terror of the desert—the blinding and stifling sand storm. Its upper boundary is rather sharply marked and often distinctly visible from any higher level. This layer includes also the city pall, that 4 tons a day, per square mile (average for Chicago, and there are worse places) smudge of soot and dirt that shuts out so much of the health-giving radiation of the sun.

- The convection layer, or stratum of diurnal convection, marked by the dust carried up from near the surface by warm ascending currents. It therefore is deepest and dustiest during summer droughts, and over arid regions. Its upper surface, perhaps two miles high, frequently is seen by the aviator or balloonist almost as distinctly as the surface of an ocean, and even to resemble that surface through the emergence above it here and there of cumulus clouds that look like so many islands. This is the next dustiest of the shells of the konisphere, but even so its burden seldom is heavy enough to bother in any way those who move about in its densest portion—at the surface of the earth.
- The tropic layer, or layer coincident with the troposphere, and therefore 6 to 7 miles deep in middle latitudes and two or three miles deeper, on the average, in tropical regions. Its top is the limit of even occasional convection, and the dust of its upper levels relatively both sparse and fine. Its upper surface rarely has been observed, since aeroplanes and balloons seldom pass that level. We know where that surface is, however, because every ascending current of air necessarily carries with it some of the dust of the lower levels, and therefore dust of terrestrial origin must extend to the upper limit of convection, that is, to the top of the troposphere, and no farther. We also have observational evidence of this upper surface through the effect of the dust particles on sunlight.
- The stratic layer, or the region of all the atmosphere of appreciable density beyond the troposphere. The dust of this region is of two parts; one roughly constant in amount, the other extremely variable. The first comes from the myriads of meteors that hourly enter the atmosphere. The second is

due to occasional volcanic explosions of great violence which, like those of Asama in 1783, Krakatoa in 1883, Katmai in 1912, and many others, hurl powdered rock far beyond the levels of the highest clouds.

The heavier dust particles of the turbulence layer quickly settle of their own weight. To a less extent that is true also of the dust in the convection layer. In the main, however, the finest particles in the troposphere are carried down by condensation, either of water vapour directly onto them or as a result of being picked up by falling drops or drifting snow flakes.

In this way the whole lower atmosphere from the surface of the earth to the tops of the highest clouds—a layer 6 to 10 miles thick—is literally washed, or scrubbed, as such processes are called, by rains and snows. If the air were so dry that there could be no precipitation it quickly would become suffocatingly filled with fine dust. In fact, it is believed by some that our sister planet, Venus, has just such a waterless, dust-filled atmosphere.

The dust of the stratosphere is not so fortunate, if we may put it that way, as that of the lower levels. It must get down through this region by itself, for there is not enough vapour up there to lend it any aid on its earthward journey. Often it is years in getting out of this arid realm, but once it has covered that part of its course the rest of the trip is quick and easy by way of the snowflake and raindrop routes. Even the dust of the earth, therefore, reveals a considerable structure of the atmosphere—at least four distinct layers. And it falls into still other great divisions according to this or that basis of separation.

Nearly all the foregoing concerns the troposphere. The little that follows relates to the stratosphere, about which our knowledge still is very slight.

Upper Trades

Since the stratosphere is much warmer, 30° F. to 40° F., in the polar regions, than in the equatorial, it would seem that there must be an upper interzonal circulation of the atmosphere somewhat like a mirrored image of the lower—towards the poles in its under portion and from them in its upper levels.

The rotation of the earth obviously would affect this upper interzonal circulation in the same manner that it does that of the troposphere, and therefore lead to east winds over the tropical and adjacent regions, and west winds over the higher latitudes. Furthermore, a little calculation based on the temperature of the atmosphere at various levels in high and low latitudes shows that this upper circulation must begin at the height of 10 to 12 miles. This calculation further shows that at that level the average pressure must be nearly the same everywhere, and therefore the average wind at this level very light. Both these conclusions, namely, that the winds 10 to 12 miles above the surface of the earth must be light, and above that level from the east in tropical and adjacent regions, are supported by all the observational data (a fair amount) we have on the subject.

Twilight Top

We see the course of a small beam of sunshine in a darkened room because there are dust particles in the air that scatter the light. That is why a few whisks over the floor with a dry broom makes the beam brighter.

This explains, too, why we can see the shaft from the searchlight, and the focusing "streams" when the sun is "drawing water." In each case the contrast is between the myriads of illuminated motes and the shaded, hence darker, portions of the surrounding air. Not only the dust particles, but also, though to a far less degree, all the gas molecules of the atmosphere, are luminous in sunshine. This air luminosity is the chief factor in the blue of the sky, and an important factor in other sky colors.

It accounts also for the twilight arch—the visible boundary between the shadow of the earth and the illuminated atmosphere—that rises above the eastern horizon as the sun sinks beneath the western.

Since the observer is himself within the earth's shadow it is obvious that by noting the exact time this arch is directly over head, say, he may know, from certain astronomical tables, just how many degrees the sun is then below the horizon; and that from this value, in turn, and the radius of the earth he can compute the height of this arch, that is, the greatest height at which the density of the air still is sufficient to scatter a perceptible amount of incident sunshine. Numerous measurements of this kind have been made, and that height thus found to be about 44 miles.

In respect, then, to its efficiency as a light-scattering agent also the atmosphere has structure, an inner shell about 44 miles thick in which the scattering is appreciable, and an outer in which it is imperceptible.

Auroral Base

The polar lights, both northern (aurora borealis) and southern (aurora australis) divide the atmosphere into distinct parts, an inner, about 62 miles deep, into which auroras generally do not penetrate, and an outer of unknown thickness, but certainly hundreds of miles, in which they commonly do occur.

Kennelly-Heaviside Layer

When radio-telegraphy over long distances was first attained we were much puzzled to know how it could be, for surely radio waves are just greatly magnified light waves, and light doesn't bend to the curvature of the earth in such manner that an object can be seen a thousand miles away. But if the air were highly transparent, and both the earth and the encircling sky excellent reflectors, then a powerful light at London say, might well be seen from New York, or any other place on the globe. The light could not get through either reflector and therefore would keep on traveling between them until finally absorbed. The same is true also of electric waves, and for these the earth is a reflector.

If therefore, the sky reflected them too we would expect long distance radio communication to be possible, but not otherwise. Hence, when such communication was accomplished, Heaviside and Kennelly told us that the sky must be a reflector of electric waves, that is, an electric conductor, and everybody answered. "Why, of course, it is a conductor."

And then came the long search to find how it is made a conductor and at what level its conductivity is adequate to account for the phenomena observed. We believe now that this conductivity is owing essentially to the presence in the upper atmosphere, 30 to 60 miles or more above the surface, of a million or so free electrons per cubic inch, due to solar radiation in the far ultra violet. The height of the under surface of this reflecting region, or Kennelly-Heaviside layer, as it commonly is called, is not sharply determined since it appears to vary with the wave-length of the incident wireless wave; nor is it constant for any given wave-length, but varies from day to night and from season to season.

But despite these variations the atmosphere, in respect to its electric state, and its relation therefore to wireless waves, always consists of two parts: a highly ionized, conducting and wave-reflecting outer shell, the Kennelly-Heaviside side; and a relatively non-conducting, but wave-transmitting inner shell. Electrically, also, the atmosphere has structure.

Ozone Layer

As explained in the chapter on the composition of the atmosphere, there is very little ozone (triatomic oxygen) in the lower air up at least to the level of the highest clouds, but certainly very much more somewhere beyond that height. From spectroscopic observations several persons have computed the height of the centre of gravity of the ozone to be 25 to 30 miles above the surface of the earth. But whatever the correct value of this Iteight, surely in respect to ozone also the atmosphere has its structure—an intermediate shell rich in ozone, and an inner one and an outer that contain practically none at all.

DISTRIBUTION OF TEMPERATURE

The pull, as we call it, of gravity makes water run down hill. It also makes a heavy liquid underrun a lighter one in the same level; both are drawn in the direction of the bottom, but the pull on the heavier, or denser, is greater than on the lighter, and the stronger pull prevails. Gravity also makes an isolated mass of liquid or gas in a heavier one go up, not down; it is pushed or buoyed up by a force equal to the difference between the weight of the lighter and that of an equal volume of the heavier.

Clearly, then, whenever two masses of air of unequal density come into free contact with each other the lighter is pushed up and away, except in the case of properly adjusted winds, as explained in the previous chapter. Now air rapidly increases in volume, and correspondingly decreases in density, with increase of temperature—roughly 1% per 5° F. at ordinary temperatures. Hence the hot air in a chimney is lighter than an equal volume of the cold air on the outside, and therefore is pushed up by the latter which, in turn, is heated and itself pushed up, and so on as long as there is a fire in the grate to supply the heat.

To be sure, the combustion alters the composition of the air (makes it richer in carbon dioxide if coal is used, and in both carbon dioxide and water vapour if wood or gas is the fuel, and poorer in oxygen) in such manner as to render that in the chimney heavier, at the same temperature, than that outside, but this increase in density through change in composition is small in comparison to its decrease in density by heating. At most it could balance or offset a temperature increase of only about 40° F. over coal, or 10° F. over wood, while ordinarily the effect if much less, since commonly only part of the oxygen is consumed; hence the heating, being several times this maximum value, has, in any case, the best of the argument, as it were, and the chimneys keep on drawing.

Similarly, air in the open is underrun and pushed up by even slightly cooler adjacent air of the same composition, unless, as already explained, the two masses happen to be flowing past each other in the right positions and directions and with the proper velocities. Actually, the heated air expands as its temperature rises, and overflows above wherever its pressure is thus made greater than that of the adjacent atmosphere. This overflow, or outflow, decreases the pressure at the bottom, and in the lower portions, of the heated air, and at the same time increases the pressure round about under the places of overflow—mass, hence weight, is removed from one place and added to others. This disturbs the balance.

Gravity tends to restore it and thereby induces winds in the direction, initially at least, of higher to lower pressure. If the heated region is very small, equilibrium is quickly established, unless the heating is maintained. But where the higher temperature covers a large the winds no longer flow directly towards the centre of lowest pressure but more or less round about it, owing to the rotation of the earth, in a manner seemingly most contrarious. This heating in innumerable cases is very local and of only a few hours' duration; in many others it is quite extensive and lasts days Weeks, and even all season long; while its greatest manifestation is the year after year and age after age continuously higher temperature in tropical realms and lower in the frigid zones.

This perpetual heating of the atmosphere over one great region, and its ceaseless cooling over another, or rather, two others, keeps it continuously out of balance and makes the winds, especially the trades and the westerlies, forever to blow—to blow dizzily over a rotating earth, and time and again violently and confusedly incident to the rapid, the all but explosive, delivery, by condensation, to a limited region of vast quantities of heat that had been slowly accumulated by evaporation from others afar off. The whole of the atmosphere to the tops of the highest clouds, that is, the whole of the troposphere, is a huge convection system, greatly complicated by the rotation of the earth and all but hopelessly confused by evaporation and condensation. The stratosphere, too, has its circulation, but as yet not much is known about it.

Of course it is difference in pressure *at the same level* that pushes the air about, or makes the winds to blow, but, as explained, this difference in pressure depends, in turn, mainly (water vapour has a little to do with it) on the distribution of temperature. That is one reason, but not the only one, why this distribution is so important. Perhaps some good physicist will insist that it really isn't difference of pressure at the same height above sea level that makes the winds blow, but difference of pressure over an "isentropic surface," or surface of "equal entropy." Well, he would be right in respect to appreciable heights above the surface, because for the free air the isentropic surface is the "level" surface. But nothing short of a surgical operation can get the idea of entropy into the other fellow's head, and there is no rivet, weld, or hermetic seal that will keep it there. Besides, commonly (not always), there isn't much difference between the two after all—"same level" and "isentropic level"—and so we will stick to the one everybody knows and no one forgets, that is, "same level."

Source of Heat

When we think of the source of heat, especially in the winter-time, we are likely to have in mind some sort of combustion, for that is the cause of the tropical climate we have indoors at that season. But indoors is a mighty small place in comparison with all outdoors; and outdoors is heated, too, often very hot in summer, and always far above the 460° below zero Fahrenheit that would be its temperature if there were no heating at all. Almost every bit of this enormous amount of heating comes from just one source, the sun.

Incoming Radiation

The radiation from the sun is so great that if it all got through the atmosphere enough would fall on each square foot directly facing it to heat a gallon of water from the freezing point to the boiling point in just three and a half hours. But it does not all get through, and what does get through always comes in slopingly except wherever the sun happens for the moment to be directly overhead. In fact, owing to the reflecting power of clouds, especially, and the surface of the earth, and to the scattering (not reflection) of light by the molecules of the air and by the myriads of dust motes, one-third, roughly, of the incoming radiation is thrown off to space without producing any effect whatever on the temperature of the atmosphere or of the earth beneath.

Another one-third, again roughly, of the incoming solar radiation is absorbed by the atmosphere, and the remaining portion by the earth. These statements apply to the earth as a whole. The ratios between loss by reflection and scattering, air absorption, and earth absorption, vary widely from place to place and season to season, owing mainly to differences in humidity, cloudiness, and elevation of the sun above the horizon, and differences in the character of the surface of the earth—whether land, water, snow or ice, bare soil or vegetation.

Clear Sky Radiations

It is interesting to note that the amount of radiation reaching the earth from a clear sky is equal to a considerable fraction of that which reaches it from the sun directly. At sea level the amount of sky radiation onto a horizontal surface of any particular size, a square foot, say, is equal to about 7.8% of the amount of unaffected, or direct, solar radiation onto an equal area squarely facing the sun at the same time and place. When the sun is directly overhead its supply of heat to a horizontal surface at sea level is nearly 13 times as great as that from the sky. When it is one-third of the way down from the zenith to the horizon its contribution of heat to the earth is only a little more than 6 times that from the sky, and each is then decidedly less than it is when the sun is in the zenith. Finally, the two sources are equal, though both are still further enfeebled, when the sun is above the horizon about one-twelfth the distance to the zenith.

The brightness of the clear sky is greatest near the sun, as even casual observations readily show, and decreases gradually with increase of distance therefrom over a large part of the whole area. Hence the total of sky radiation received per minute on a horizontal surface is greatest at noon, as is also the direct solar radiation.

The intensity of sky radiation decreases, in general, with increase of height above sea level, while that of the direct solar radiation increases.

RADIATION FROM AN OCERCAST SKY

When the sky is overcast, neither its brightness nor the total amount of radiation received on a given horizontal area is at all constant, even for the same height of the nun, because the clouds in question may be of any kind from the thinnest cirrus that just dims the sun, to the darkest nimbus that reduces even noonday brilliance to twilight. If, however, the sky is completely overcast by an approximately uniform cloud layer dense enough, but not greatly more than enough, to prevent the position of the sun from showing, then the total radiation from this cloud layer onto a horizontal surface is, on the average, slightly greater than that from a clear sky. Evidently, too, the amount received of this cloud-transmitted radiation generally must increase with increase of height above sea level of the place of reception. The brightness of the cloud layer is surprisingly close to uniform. It is greatest nearly overhead (just a little way off in the direction of the sun), but still nine-tenths as bright half way to the horizon, and half as bright almost at the horizon.

DISPOSAL OF RADIATION OF SURFACE OF THE EARTH

There are just three things that can happen to radiation incident onto any extended object. It must be reflected, transmitted or absorbed. When the object is extremely small it more or less scatters incident radiation, and radiation that just grazes the boundary of an object suffers still another effect which we call diffraction. However, neither scattering nor diffraction occurs when the object is large and its edges are not involved. They do not, therefore, occur in the case of radiation incident on the surface of the earth. Here then, the incident radiation is all used up by two processes, reflection and absorption, since there is no transmission—no passage of radiation through the earth and out at the other side. The portion reflected is about 70% for snow-covered regions, and 7% for the rest of the world. The remainder is absorbed, that is, 30% wherever there is snow, and 93% at all other places, both land and water. That which is reflected is lost except in so far as it is absorbed by the air above. The absorbed portion goes largely to heating the upper layers of the soil or water, but not all of it, since a considerable part is consumed in maintaining evaporation, and a much smaller part in effecting plant growth and development. Another relatively small part merely melts snow and ice without raising their temperature above the freezing point.

The heated surface in turn heats the soil or rock by conduction, but appreciably to a depth of only a few feet. The heating of water extends to a greater depth owing partly to the penetration of the rays to some distance below the surface, and partly to the mixing of the water by wave action. The heated surface also warms the air above it both by direct contact and by radiation. Furthermore, the heated air through convection shares its warmth with other and colder air above; and the heat consumed in evaporation at one place is liberated, that is made sensible or temperature-producing, some other place, usually in mid-air, where condensation occurs, and far away.

Practically every bit of this heating of earth, ocean and air, and supply of energy for evaporation, plant growth, ice melting, and what not else, comes from the sun—all directly except about one part in half a million that reaches us after reflection by the full moon and the planets. A negligibly small amount comes from the fixed stars, enough to keep the average temperature of the out-doors air about two millionths of a degree Fahrenheit higher than it otherwise would be. Finally, another very small amount comes from the heated interior of the earth.

QUANTITY AND EFFECTS OF HEAT FROM THE INTERIOR OF THE EARTH

If the earth had no atmosphere, and if there were no sun or stars to send us a flood of radiation, the supply of heat from the interior (of which four-fifths, roughly, is from radioactive material) alone would keep up the surface temperatures to about 60° absolute, on the Fahrenheit scale, that is, -400° F., approximately. Hence the flow of heat from the interior of the earth per square foot of surface is sufficient to raise the temperature of a gallon of water about 1° F. in 16 days, and the total flow through the whole surface out to space sufficient to heat 92,000 tons of water from the freezing to the boiling point every second of time; or enough, starting at room temperatures, to melt 1,000,000 tons of lead per second.

These are big figures, and yet all this flow of heat keeps the actual temperature of the surface of the earth only about 1/25 of a degree F. higher than it otherwise would be. The figures also tell us the surprising story that if 10,000 times as much heat came from the interior of the earth as now actually does come, or, what amounts to the same thing, if everywhere there was a sea of molten cast iron covered over with a layer of rock and dirt only 10 to 12 feet thick, the oceans above could rest thereon serene with no close approach to the boiling point, so excellent an insulator, or poor a conductor, is this material; and that if the dirt and rock crust were 20 feet thick we could go about over it ourselves in perfect comfort!

Outgoing Radiation

On the average, the earth loses to space, or emits to space, by radiation very approximately the same amount of heat each year that it absorbs of incoming radiation during the same time, plus, of course, the supply of heat that reaches the surface from the interior. We know that the loss is substantially equal to the gain because otherwise the surface would be growing warmer from year to year, and we know that this loss is by radiation as there is no other way for the loss to occur—there being no such thing as conduction to empty space. The amount of this loss, or radiation of the entire earth to space, can be estimated from our knowledge of the incoming radiation and the fraction of it that is ineffective through scattering and reflection, especially by clouds.

It can not be measured directly because we have no means of getting out beyond the atmosphere and from that ideal place pointing our heat-gathering apparatus towards the earth. But, as implied above, we can make a pretty close estimate of the average rate at which radiation is going out from the earth as a whole, and the conclusion is that it is very nearly the same as that from a perfect radiator, or "black body," at the absolute temperature 454° on the Fahrenheit scale, or -6°

F. This is sufficient, per square foot of surface, to heat a gallon of water from the freezing to the boiling point in about 20 1/2 hours.

The radiation from the surface of the earth often is very much greater than this value, even twice as great, or more, because the temperature of the surface frequently is far higher than the -6° F., here assumed. On the other hand, at times and places, owing to very low temperatures, it is much less. On the average, however, the radiation from the surface is much in excess of that which finally gets away to space-greater by the amount of return radiation it absorbs (nearly all of it) from clouds and the atmosphere.

The surface of the earth radiates at a relatively high temperature, hence in comparative abundance. Some of this radiation goes directly through the atmosphere, but ordinarily most of it is absorbed by the water vapour and clouds in the lower air, and a little by other things, especially ozone (when the sky is clear, for it is above all clouds) and carbon dioxide. That which is absorbed in the lowest layers is, in general, reradiated, but at a lower temperature than that of the surface. This reradiation is in every direction, half of it downward, some of which is absorbed on the way, and the rest by the surface whose initial temperature and radiation it thus helps to maintain; and half upwards to the next higher layers; and so on up and up from layer to layer, but always with decreasing absorption by the air still above and increasing absorption by that below until the entire atmosphere is left behind.

Clouds do not Check Radiation

It is a well-known fact that during still clear nights the surface of the earth, and, through it, the adjacent air, cool to a much lower temperature, especially over level land and in valleys and bowl-like depressions, than they do when either the sky is clouded, or the wind is strong. Furthermore, the lower the clouds, other things being equal, the less the cooling. This interesting and important fact is clearly "explained" in many elementary books and numerous articles on the assumption that clouds and winds check the radiation of the surface of the earth, that is, make it radiate more slowly.

That seems very simple, and would be but for one little fly in the ointment—there isn't a word of truth in it. It might do perhaps as a dose of mental paregoric for a kid with the quizz colic, but it is no good for anything else. Just one thing alone ever reduces radiation, and that is decrease of temperature. They are, however, themselves good radiators; and as their temperature, when they are low, is nearly that of the earth, they send down to it almost as much radiation as it itself emits, and as practically all this cloud radiation is absorbed by the earth, it follows that the surface temperature remains substantially constant. It isn't that the radiation from the earth is checked in the least, but that it receives from the cloud canopy and absorbs wellnigh as much as it itself gives out. Neither is the approximately constant temperature maintained by an appreciable wind owing to any check whatever exerted by it on surface radiation, but to the fact that the net loss of heat thus sustained, and on clear nights it is considerable, is distributed by turbulence through such a deep layer and great quantity of air that the fall in temperature is small even when the total loss of heat is large.

Temperature of Surface Air

When we talk about the "surface" air it often is advisable to explain just what air we have in mind, for this term is quite flexible. We might mean only that air which is in actual molecular contact with the surface, or that which at most is within a few inches of it, or finally, all below the height of eight or ten feet, the air to which we chiefly are exposed while outdoors.

The temperature of the surface air, in any one of these senses, is determined mainly by that of the surface itself. Whatever the temperature of the surface, that also is the temperature of the contact air, and very nearly the temperature of all that air which by turbulence or otherwise is frequently brought into contact with the surface. On the stillest of nights this layer at places may be only a few thick. During the daytime, however, especially when there is sunshine to induce thermal convection, and whenever, day or night, there is a measurable movement of the air, it is certain to be at least a good many feet thick.

The essential point in this: The temperature of the air near the surface (how near varies with the circumstances) depends more on contacts with that surface than it does on the amount of solar radiation to which it may be exposed. The surface temperature of course does vary with the intensity and duration of the sunshine, and so therefore does also that of the surface air, but indirectly through contact with the surface and not directly by absorption of solar energy.

Relation between Surface Temperature and Temperature of Surface air

As above stated, the temperature of the actual contact air must be the same as that of the surface (of the substance, ground or what not, at its surface) against which it rests. If this surface air remained fixed, as we often are told that it does, then it would seem that the air next in contact with it also should become fixed in position, and so on indefinitely. But we know that fixity of position of the air molecules does not extend to a measurable distance from any solid, for we can blow smoke past it and see the motion of the air. We therefore are forced to the conclusion that fixity of position does not apply, at least not for any appreciable length of time, even to the contact molecules. The way out of the difficulty appears to be this: The actual contact molecules of the air are at rest, like a liquid film, but they do not stay at rest. They evaporate, and as they leave the surface others condense thereon—are adsorbed—a continuous process the details of which are not yet all known. In this way the contact molecules, during the extremely brief interval of their contact, are fixed in position, but they are continuously reverting to the gaseous state, and therefore the atmosphere at ordinary temperatures is always fluid however measurably near it may be to the surface in question.

Since the air is directly heated chiefly by contact with the surface of the earth, and indirectly by the sharing of this heat, through convection, with colder air above, it follows that in general wherever the temperature of the lower atmosphere is increasing, that is, over nearly all snow-free land, and particularly during the day time, there the average temperature of the surface is higher than that of the free surface air. This is in accordance with what physicists call the second law of thermodynamics, and what everybody else knows without calling it anything, namely, that the temperature of the heater is higher than the temperature of the thing heated. Similarly, where the lower air commonly is cooled, as it is over snow-covered regions, there the average temperature of the surface is lower than that of the surface air—the cooler is colder than the thing cooled.

Maximum and Minimum Temperatures

Obviously if the heater changes temperature, so also will the heated, and the heater will be the first to change and the first to reach its extreme values—maxima and minima. There is no surprise, therefore, in the fact that the daily maximum temperature of a snow-free land surface occurs earlier, about 1 o'clock P.M., than that of the air above it, which is delayed until around 3 o'clock.

The air and surface minima, occurring near daybreak, are much closer together, owing partly to the slow cooling of the soil through the night. Over the ocean the temperature of the air normally is a little higher, a degree or so, than that of the water, and the time of its maximum value, near 1 o'clock P.M., a little earlier than that of the water. This is due to the fact that here the surface air is humid and also "dusty" with salt particles and therefore absorbs a large amount of radiation, so much indeed that its daily range of temperature is more dependent on this direct absorption than it is on conduction and convection from the surface. The minimum temperatures of air and water occur simultaneously, or nearly so.

Periodic Temperature Changes

Nearly 150 different periods of temperature and other weather changes have been published, ranging from 24 hours to 744 years. Nearly all of them, however, have a shorter period than 40 years, and half of them a period of 8 years or less. Of this great number of periods there are only two, the 24-hour or daily period, and the 12-month or annual period, that everybody accepts.

There is one other, the so-called 11-year or sunspot period, that is widely, though not universally accepted; and still another the 35-year, or Brückner, period that many believe to be real. No credence was ever given to any of the others save perhaps by their discoverers, and in most cases even that must have been half-hearted. The daily period is everywhere conspicuous (save for part of the time in polar regions, when the sun is continuously above or continuously below the horizon), and in respect to temperature, gives, on the average, a maximum in the early to mid afternoon and a minimum shortly before sunrise.

Over the oceans this diurnal range is only 1° F. to 3° F. as a rule. It also is small in the humid and cloudy portions of the continental tropics, owing to the large amount of return radiation from the clouds and water vapour. In desert regions, especially at high altitudes, where the sky is clear and the humidity very low the diurnal range of temperature is at its maximum. In extreme cases this range is of the order of 100° F., from distinctly below freezing to decidedly over 100° F.—both in the shade. The annual range also is extremely conspicuous in most parts of the world.

In this case the exception does not occur at and near the poles, but at and for some distance on either side of the equator. At the equator the "year," as it were, counted from the time the sun is overhead at noon until farthest away (23 1/2 °), and then back again is 6 months, not 12. Next beyond the equator on either side there obviously are two such "years" but of unequal length. At one distance they are 5 and 7 months, at another 4 and 8, and so on until at the Tropics, Capricorn and Cancer, only one is left, and that one 12 months in duration, the same as from there on to the pole.

The times of occurrence of the annual maxima and minima vary widely from place to place, but always they are after maximum and minimum reception of heat from the sun. The delays are least over inland deserts and greatest over mid to high latitude portions of the oceans.

The sunspot period, approximately 11.1 years, is most pronounced at high levels within the tropics. Here the average temperature during the year or two around sunspot minima, or when the spots are fewest and smallest, is about 2° F. higher than the average temperature during the time of spot maxima. The same relation appears to hold, in general, for the middle and higher latitudes but with decidedly less contrast.

The Brückner period is very irregular in length, varying from roughly 20 years to perhaps 50, and the amplitude of its temperature range uncertain but always small. Its irregularity in length deprives it of practically all forecasting value, and indeed makes its very existence as anything other than a fortuitous recurrence highly doubtful.

There are two other known and real periods in respect to average temperatures and other climatic elements, but they are far too long to consider in any business affairs. One concerns the slow change of the season of the year when the earth is nearest the sun, due to the combined effect of the motion of the perihelion and the precession of the equinoxes. Just at present the earth is nearest the sun the first week of January and farthest away the first week of July; and this difference in distance is sufficient, if long continued, to vary the average temperature of the earth by at least 7° or 8° F.

That is, at present the winters of the northern hemisphere are shorter and milder, and the summers longer and less hot, than they would be if we were nearest the sun the first week of July and farthest from it the first week of January, as we were about 10,500 years ago, and, in the same length of time, will be again. This is one reason, and the unequal distribution of land and water another, why the average temperature for the year is about 2° F. higher in the northern hemisphere than in the southern, and why the thermal equator is north of the geographic equator.

Beyond question this particular period is of great climatic importance, but we know all about its course and its cause and the changes it effects come about so slowly that practically they do not concern us at all. The other period referred to is that of the changes in the ellipticity of the earth's orbit or variations in the difference between the annual maximum and minimum distances of the earth from the sun. But the length of this period, roughly 100,000 years, keeps it out of every business equation however prudently constructed.

Just to make the list complete it may be worthwhile to mention a few utterly unimportant but entirely real temperature periods. Those are the periods of the changes in light and heat received from the moon-maximum at full moon, minimum at new moon; changes in the distance of the earth from the sun due to the pull of the moon in its orbit about the earth; and similar but far less changes of and by the planets. The sum total of the effects of all the planets is about equal to that of the moon alone, that is, a change in the average temperature of the earth of about.02° F., due almost wholly to variations in our distance from the sun, or, as we say, to perturbations in the earth's orbit. But, as already stated, this change in temperature is too small to bother about.

Temperature Lag

It was stated above that the hottest time of the day is not noon, when the sun is most effective, but two to four hours later; and similarly, that the coldest weather does not come with the shortest days, but generally a month or so later. In proverb form: "As the days grow longer the cold grows stronger." In the early morning of a clear day following a cloudless night, say, the earth and surface air are relatively cool. Then with sunrise they begin to warm up, but not rapidly, even when there is no wind, because it requires an appreciable amount of heat to warm even a pound of soil 1° F., and several times as much to equally warm a pound of water.

But as the sunshine continues, the soil at first gets hotter and hotter, and as its temperature rises the rate at which it loses heat by radiation rapidly increases. However, since the soil, including of course its covering, warms slowly, owing to its large capacity for heat, its loss by radiation falls more and more behind its gain by absorption as the sun rises higher in the heavens, and therefore catches up with the latter only in the afternoon when the insolation is distinctly less than it was at midday.

Hence the diurnal maximum temperature, whether of the lower air or of the surface of the earth (an earlier phenomenon) necessarily lags behind the maximum intensity of the sunshine. Similarly, the annual maximum temperature occurs several weeks after the days are longest and the heating strongest. Very similarly too, because the earth can give off stored up heat when the supply becomes deficient, the minimum temperature comes several weeks after the shortest days. During this period, as the days grow longer the cold grows stronger.

The diurnal and annual heating and cooling, and lagging of temperature extremes, may be likened to the alternate rise and fall of the water level in a reservoir having a continuously open drain pipe at the bottom and a periodically variable inflow, now greater, now less, than the then rate of outflow, but so regulated that the reservoir may never become empty.

Day Degrees

Not only are we interested in the values and times of occurrence of maximum and minimum temperatures but also, and even more, concerned in the occurrence of certain critical temperatures. For instance, we are very much interested in the temperature at which frost can occur until it does occur, after which, if it has been a "killing" one, we are no longer much concerned as there is nothing left for the next one to injure.

Another critical temperature is 42° F. as that closely marks the boundary between growth and dormancy for most vegetation of the temperate zones. In fact it is customary to call the difference between the average temperature of a given day, if higher than this value, and 42° F., its day degrees. The sum of these daily values over a week, month or season, is the number of day degrees for that period, and is an important index to what might have been the vegetable growth during the time in question. Similarly, engineers and others interested in artificial heating of buildings, count day degrees relative to a temperature of 65° F.

Occasional Extremes

Once in a while an exceptional combination of conditions brings to a given place an abnormally high or low temperature, usually for only an hour or two, or a day at most, but sometimes for several days together, and even a month or longer. It is always easy to know from the current maps of weather distribution exactly what caused the extreme in question, but it never is possible to trace them farther back than two or three steps at most, nor very long to foresee their coming. Some of them one never forgets, and a few continue for a century or more to put disconcerting humps or depressions on our statistical curves.

Wind Direction and Temperature

The effect of wind direction on the temperature of a place depends on its location. Well within the Tropics, and also near the poles, the effect of wind direction obviously is small because the temperature is pretty nearly the same round about in every direction. In middle latitudes, however, the situation is quite different, partly because here the temperatures commonly are not the same in every direction, and partly also, in fact mainly, because here each section of the cyclone and of the anticyclone has its own wind direction, and some of them a wind system entirely distinct from that of the others. In the forward or eastern portion of the anticyclone the winds are from the region of higher latitudes, and having come a long ways often are distinctly cool to cold for the place and time of year.

Similarly, the winds of the western segment, having come from much nearer the equator, usually are relatively warm. In the cyclone, or widespread disturbance, all that segment of 90°, more or less, lying between one line running east, to southeast, from the storm centre and another generally south to southwest (in the northern hemisphere; east to northeast, and north to northwest, in the southern hemisphere) is occupied by a great current of warm air from low latitudes. The rest of the storm area is covered with cold winds from the east, north, and northwest, in succession as one in the northern hemisphere passes from the front to the rear of the storm centre on the poleward side; from the east, south, and southwest, in the southern hemisphere.

In general, all these cold winds in each hemisphere are of polar, that is, high latitude origin. Clearly then, the temperature of the air in a cyclonic region is likely to change with the direction of the wind. In one portion of this disturbance, namely, along a narrow strip that meteorologists call the cold front, or wind shift line, and which commonly runs west of south (west of north in the southern hemisphere) from the storm centre, the wind direction rapidly changes from southwesterly to northwesterly, with, as a rule, a sharp drop in temperature as the tropical breezes give way to polar blasts. Hence in middle latitudes air temperature is closely dependent upon wind direction, both in cyclones and anticyclones; and that means the greater portion of the time, for usually we are in the midst of one or the other of these disturbances.

LOCATION AND TEMPERATURE

Every one knows that the average temperature of the tropical regions is higher than that of the polar areas, but it is not a familiar fact that, nevertheless, int he course of a year the temperature reaches 90° F. or over (in the shade) on more days in central Alaska than at Panama. It also is a surprising fact to most of us that the average temperature through January at St. Louis is the same as that in southern Iceland; and that the average temperature for the entire winter, December, January and February, at Sitka, Alaska, is about the same as that at Washington, D. C. And there are lots of other similar surprises, as, for instance, the fact that semi-tropical vegetation that would be killed by frost is now growing wild on the Scilly Islands in the latitude of northern Newfoundland.

Clearly, then, difference in longitude may be accompanied by nearly as great a variation in temperature as is difference in latitude. During winter especially the lines of equal temperature run far poleward over the oceans, and equatorward over the continents.

Another matter of great influence on the local temperature is the nature of the surrounding area, both near and distant. A fair inland point, for instance, becomes much hotter in summer and greatly colder in winter than does a mid-sea island at the same latitude. Also a coast where the prevailing winds are on-shore has a more nearly even temperature than one of the same latitude with off-shore winds. The first is bathed in ocean breezes of relatively equable temperature; the second in winds that have traveled far over land and that therefore are characterized by its temperature irregularities and extremes.

It is interesting in this connection also to note that the temperature is a little higher, and in some cases quite noticeably higher, in a city than in the adjacent country; and further that in the country the forest is cooler in summer, and slightly warmer in winter, than the open fields.

In some cases the foot of a high mountain occasionally is very peculiar in respect to temperature. It may happen that the air is quite cold when all of a sudden there comes across and down the mountain a roaring hot wind that within a few hours clears away every trace of even a deep snow. This wind went up the other side of the mountain saturated and rainy, hence it cooled relatively little with ascent owing to the latent heat of vaporization there rendered sensible by condensation. As it came down, however, it was dry and therefore heated rapidly with descent and consequent increase of pressure. Such are the famous foehn winds on the northern side of the Alps, and the chinook winds of the Rocky Mountains.

South Pole Colder than the North

We often are asked which is the colder, the north pole or the south. One answer might be that, as we have no records longer than a few hours at either, we don't know. That is true enough so far as bare statistics are concerned, but in this case we can reason the matter out from sure and simple premises: (*a*) Normally, the greater the height the lower the temperature, other things being equal; (*b*) the south pole is 10,000 feet, roughly, above, and the north pole at, sea level. Hence we should expect the south pole to be the colder of the two. (*c*) The greater the height the less, in general, the amount of cloud, water vapour and other gas to radiate back to the earth and thereby help to keep up its temperature; (*d*) the south pole is much higher than the north.

The south pole, therefore, because it gets less return radiation than does the north, should have the lower temperature. (*e*) The faster heat is supplied by the surface to the lower air the higher the temperature of that air, other things being equal; (*f*) by actual measurement the ice over the Arctic Ocean gives off enough heat in 24 winter hours, coming from the relatively warm water below, to increase the temperature of a layer of air 450 feet thick by 20° F., while that given off at the south pole is only a small fraction of this amount, owing to the much poorer conductivity of the snow and the far greater depth to a temperature equal to that of the arctic water. Then for this reason also the south pole must be colder than the north. It seems therefore that even without the actual observations we may feel reasonably certain that the temperatures of the two poles are not the same, especially their night temperatures since these occur when they can be but little affected by the variations in our distance from the sun, and that the south pole is distinctly the colder of the two.

Relation of Surface Temperature to Height

Those who live in mountains regions know by personal experience that, in general, the higher up the mountain the lower the temperature. Instrumental records show that this relation is true not only for mountains but also for hills and even plateaus, and that the approximate numerical values are 1° F. decrease per 330 feet ascent on a mountain, 365 among hills and 455 on plains.

The reason for this difference in favour of the plateau is the fact that the air is heated mainly by the surface of the earth which, in turn, is heated by the sunshine. That is, in the case of the plateau the surface which is the heater of the air is at the height in question all around as far as the level area extends, while the air on the mountain is affected in part by the temperature of the free air, especially when there is an appreciable wind, and this free-air temperature is lower, as we shall see presently, except on still clear nights, than surface air at the same altitude.

Relation of Temperature to Height in the free air.

According to observations the temperature of the air normally decreases with increase of latitude from the surface of the earth up to the height of several mile, roughly 6 to 7 in middle latitudes. Beyond this level the temperature remains substantially constant up to the greatest heights yet attained, probably around 18 miles. The average rate of decrease of temperature of the free air with increase of height is about 1° F. per 300 feet, from the surface up to the level at which appreciable decrease ceases—up to the "tropopause," or limiting reach of convection. In general this temperature decrease is most rapid in the upper half to two-thirds of the depth under consideration and least rapid in the lower one-third. Immediately above a land surface the change of temperature with height is widely variable, from a *decrease*, perhaps ten fold the above average value, to an *increase* ten fold that rate, at least through the first 20 feet or so, above suitable regions, such as plains, valley bottoms and bowl-shaped depressions, and, of course, during still clear nights. Few phenomena of the atmosphere are as often "explained" as is the fact that, in general, temperature decreases with increase of height, and hardly any other as inadequately, not to say erroneously, explained. The facts are:

- Half, roughly, of the sunshine that is not lost by reflection or by scattering, that is, half of it that does any heating at all, gets entirely through the air and is absorbed by the surface of the earth, which, on being thus heated, heats in turn the adjacent or lowest air.
- The other half of the effective radiation from the sun (portion used and not immediately lost) is absorbed mainly by the water vapour, and as the density of this vapour generally decreases very rapidly with increase of height it follows that by the direct absorption of sunshine the heating of the air likewise is greatest in its lowest levels.
- The surface of the earth loses heat (it doesn't keep on getting hotter and hotter indefinitely) not only by conduction to the adjacent air, but also by radiation, a kind of radiation greatly absorbed by water vapour. Therefore in this third way, too, as in each of the others, the atmosphere is more and more strongly heated with decrease of height.
- Although the surface air is most heated it is not equally heated everywhere. Hence the warmest and lightest portions are pushed up—forced to rise—by the cooler and denser air round about as a cork is bobbed up when let go of under water.
- The ascending (pushed-up) warm air comes under less and less pressure with increase of height by the weight of the air left below. It therefore continuously expands while rising, and all the time against pressure—the weight of whatever air is still above it. But this expansion against pressure is work,

and work at the expense of the only supply of energy the rising air has—its heat. Hence as it rises it must and does become cooler.

- This cooling of the air by convection does not go on to absolute zero, nor, as explained, does the surface air keep on getting hotter and hotter. And these limitations are owing to the fact that the atmosphere loses heat by radiation. The free air therefore is all the time gaining heat by absorption of radiation and losing heat by emission of radiation, while the surface air is gaining heat also by contact with the warmed earth.
- But loss of heat by radiation decreases very rapidly with fall of temperature, while the power to absorb radiation does not change. Hence as the air ascends higher and higher, and thereby gets colder and colder, it presently comes to a temperature at which its loss by radiation is equal to its gain by absorption. Beyond this level it can not rise, because if it did so it instantly would become colder and denser than its environment and fall back again. From this level on up the temperature of the air must remain roughly constant except as modified, perhaps, by change of composition. This is the isothermal region, or stratosphere, which every planet must have whatever the extent and composition of its atmosphere.
- Since the cold air of the stratosphere is losing heat by radiation at the same rate that it is gaining it by absorption it follows that the warmer air of lower levels is losing by radiation faster than it is gaining by absorption, the net difference at each *level* (not moment to moment for the moving air) being made up by heat brought there by convection from the surface, either directly as such, or indirectly through evaporation and subsequent condensation. Thus the whole of the troposphere, or, in other words, the whole of that portion of the atmosphere in which clouds can and do occur, is continuously being heated below and cooled above. In this way, that is, by heating below and cooling above, convection and, in general, a decrease of temperature from bottom to top of the convective layer, is maintained without the air as a whole getting either warmer or colder.

Temperature Changes in the Stratosphere

Although the stratosphere, that portion of the atmosphere beyond the highest clouds, has no immediate contact with the warming and cooling surface, nevertheless its temperature at any given locality often varies 10° F. to 20° F., and even more, not vertically, as a rule, but horizontally, or from day to day. Thus the temperature of the stratosphere commonly is distinctly higher over the forward and central portions of a cyclonic area than it is over the corresponding portions of an anticyclone. The cause of this change is not definitely known.

Temperature Inversions

As explained in the chapter on the structure of the atmosphere, during still clear nights the surface air becomes so cold over level land areas, in valleys and in basins, that often there is a rapid rise of temperature with increase of height through the first 10 to 100 feet, or more. This is the surface inversion, so favorable to production of frost, and without which orchard heating commonly would be unnecessary and, moreover, in general impracticable.

As also explained in that previous chapter, every wind of appreciable strength so thoroughly mixes up the lower air by turbulence that the temperature of the top portion of the agitated stratum is decidedly lower than that of the undisturbed air immediately above it. This is the turbulence inversion, which, because of its low temperature often is accompanied by a stratiform cloud.

It might seem that a similar inversion should occur at every interface between over- and under-running air currents, but such inversions, so far as they exist at all, are too slight to be of any particular importance. This is owing to the smallness of the friction between free air currents and the consequent all but complete absence of turbulence.

The highest temperature inversion in the atmosphere, of which we have any actual and direct measurement, is that at the base of the stratosphere during the passage of an anticyclone. As already stated the cause of this particular, and often very pronounced, inversion is not yet definitely known.

THERMAL BELT, OR GREEN BELT

There are various ways of protecting fruit from frost, but the best of them all is the proper selection of the orchard site. In a hilly or mountainous region that best location is neither on the floor of the valley nor, generally, on the top of a ridge, but, as a rule, some distance, not too far, up one or the other of the slopes. A strip along a hillside or mountainside at this level is known as the thermal belt, green belt, verdant belt, frostless belt, etc., because on still, cloudless nights this level is warmer than any other above or below it, hence least likely to have frost, and most likely to show green and uninjured vegetation.

After sundown, when the sky is clear and there is no wind, the surface of the earth everywhere cools much more rapidly than the free atmosphere, and in turn correspondingly chills the nearby air either directly through actual contact, or indi- rectly by turbulence mixing with that which had been so chilled. On the side of a hill, then, this air, because it is denser (being colder) than that at the same level over the adjacent valley, flows down slope much as would a sheet of water. However, as this drainage air reaches lower levels it evidently is subjected to increase of pressure to the extent of the weight of the air passed below.

It therefore is compressed—work is done upon it—and its temperature made higher than it otherwise would have been. In the early evening, and well up on the hill, this heating of the down flowing air causes it to become warmer and warmer with descent, almost to the valley bottom. Here the gain of heat through increase of pressure not only makes up for all that was lost by contact with the cold surface, but actually warms the air to a higher temperature than it had before it was first chilled. But this heating is pronounced only where the descent is rapid.

Air already at the bottom of the hill is not thus heated. It does not come under any greater pressure for there is no lower place to which it can rapidly drain. It therefore just gets colder and colder as the surface temperature continues to fall incident to the net loss of heat by radiation. It also gets colder, but not so rapidly, near the bottom where the slope is gentle and the current sluggish; and less and less rapidly with increase of height and gain in the speed of flow. There must therefore be some level along either valley wall at which, for the time being, the heating of the descending air by compression is just equal to its cooling by contact with the chilled surface.

Above this level the air evidently gets colder with ascent and below it colder with descent. During most of the night the flood of cold air grows gradually deeper, carrying the level of maximum temperature—the level of its crest-higher and higher up the valley sides. Near morning, however, it becomes practically stationary, and where it then is frost obviously is least likely to occur. This is the place to plant our orchard—here along the thermal belt where the temperature is highest and the chance of frost the very least; where vegetation may pass through the night unharmed, while all above is frozen stiff and all below white with frost.

Chapter 3

Water Vapour in Atmosphere

It long has been recognized that the first thing to do in cooking a rabbit is to catch the rabbit. Similarly the distribution of water vapour requires that first there shall be water vapour that may be distributed. And this raises questions about that water vapour. Where does it come from? How much is there of it? Where does it go to? How does it get there? Well, no matter what was the origin of the first water vapour about the earth, nearly all of it that at any time is now in the atmosphere got there as a result of evaporation, mainly from water surfaces, snow and ice sheets, damp soil and growing vegetation.

EVAPORATION

In picturing to ourselves the mechanism of evaporation, it is convenient in respect both to gaseous and liquid water (ice presently will be mentioned separately) to think of the ultimate particles or molecules as moving about among themselves along exceedingly short paths (except only when the vapour and all its accompanying gases are together extremely rare) in every which direction and with widely varying speeds with reference, in the case of each particle, to some point in the direction of its path and which we regard as fixed in position. This may sound a little finical, but it isn't, for really we must have something to tie to, as we say.

If we measure the velocity of anything it must be in terms of its rate of approach towards, or departure from, something else that for our particular purpose we can consider as at rest, whatever the motion of the thing "at rest" may be in respect to a third object, or of that in turn to still another, and so on indefinitely. For instance, one may consider his face as being at rest while his hands are in motion with reference to it as he eats in a dining car; or the car at rest and himself, face and all, in motion as he comes in or goes out; or the ground at rest while the car is in motion over it; or the sun at rest while the earth has several kinds of motions; or the stars at rest while—but there is no end to this and we would better stop while stopping is good.

The point is this: If we talk about velocity we always must think of, or measure, that velocity in terms of rate of movement directly towards, or directly away from, something else, which for our then particular purpose we may regard as being fixed in position. If this little digression has helped us to think straight about "velocity," so that we know what we mean, we then may return to our consideration of the linear movements (they also spin) of water molecules. In the case of an open vessel of water, for example, the sides and bottom evidently may be regarded as at rest with respect to the moving molecules, and so too may the surface of the water, if it is not agitated. The same also is true of reservoirs, ponds, lakes and other bodies of water.

As just explained, we think of the molecules of water, and also of those of every other sort of liquid or gas, as being continuously in a state of utterly confused migration both as to direction and speed. Furthermore, the greater this average speed the higher the temperature. This is the right way to look at it, but if we wish to do so, and it often is convenient, we also may say, the higher the temperature the greater the average speed of the molecules.

Now myriads of molecules at and next to any free water surface, for instance, are moving outwards, and many of them with velocities so great as to carry them into the space beyond and away from that restraining force, whatever it is, that tends to keep them within, and a part of, the body of the liquid. Outside the liquid the space per molecule is greater, in fact greater without limit, except as restricted by enclosing walls, as in a closed vessel, or by the force of gravity. Hence its free path, the distance it travels from collision to collision, also is vastly greater outside than within. All restraining force of cohesion then is gone and the swarm of free particles is a gas, or, in this case, to be more exact, a vapour, at least at ordinary temperatures—a gas which, at its current temperature, can be compressed to a liquid.

Since great velocity of molecules means high temperature, and small velocity low temperature, and since the greater the velocity of a molecule in a liquid the better its chance to get away from cohesion restraint and become a gas molecule, it follows that evaporation, the changing of a liquid to a gas, decreases the average velocity of the liquid molecules—the faster getting away while the slower remain behind—and hence lowers the temperature of the remaining liquid, or at least keeps it from getting as hot, or hotter as fast, as it otherwise would.

Not only do the molecules in liquid water get away in great numbers and become gas molecules, but also many such gas molecules, in the course of their randomly directed flights, plunge against every water surface with such velocity that they merge with and become a part of the liquid. This is condensation, the reverse of evaporation, more fully explained later. Both processes occur simultaneously, while the circumstances of temperature and pressure determine which shall be the faster of the two. By "evaporation" we generally mean net evaporation, or excess of evaporation over condensation; while by "condensation" we commonly mean net condensation—excess of condensation over evaporation.

Perhaps a duly inquisitive reader may note that nothing at all definite has yet been said about the nature of the forces that keep a liquid just what it is—a liquid—and prevent it from exploding violently into a gas. To be sure a liquid can be made to explode by heating it to a high temperature under pressure and then suddenly removing that pressure. But under ordinary circumstances a group of liquid molecules does not explode into a group of gas molecules of the same kind, and the problem as a whole still challenges the guidance of clever theories and the test of ingenious experiments.

Although, and confessedly, we soon run out of satisfactory answers to reasonable questions about evaporation we know that it does occur. We know, too, referring now to water, in which we are here particularly interested, that the higher the temperature of the water the faster it evaporates; that the lower the vapour pressure just over the water surface, and therefore also the greater the wind, since it in turn, by remov- ing the accumulated vapour, lowers that pressure, the faster the evaporation; that the greater the extent of the surface the greater likewise, but not in the same proportion, the total amount of water evaporated in a given time; that the saltier the water the slower it evaporates; and that this and that other condition also has its own effects, but generally negligibly small in comparison to those just specifically mentioned.

Although the evaporation of water has been the subject of many studies, as might be inferred from the above list of facts about it—the subject indeed of hundreds of investigations—no simple and inclusive relation has been found between the evaporation from an open body of water and the different factors that control it. Several empirical equations have been devised to express these relations, but they are far from being satisfactory.

Indeed it seems that the true relation probably is much too complicated to admit of expression in the terms of a simple equation. The best we can do, perhaps, especially in connection with climatological studies, is to obtain evaporation measurements from water pans of the same size and like exposure at favorable localities.

Such data do not tell the whole evaporation story, but they have the great merit of being both easily obtained and intercomparable, and they do give a fair idea of the relative evaporativities of the places at which they were obtained.

As stated above, ice in any form, from snowflake to glacier, evaporates too—sublimes if one insists on using the proper technical term for changing from a solid to a gas without going through the intermediate stage of being a liquid.

But it isn't easy to form a satisfactory conception of just how ice, or any other solid, does evaporate. It is agreeable enough to think of the molecules of a gas as darting in every direction among themselves like a swarm of bees on the wing, only more so and altogether blindly; and we can, without shock, liken the molecules in a drop of water to an excited school of minnows, but the molecules of a solid have no such freedom.

They are as fixed in position as barnacles on a rock. Even the atoms, that together make the molecules, keep to their respective places, as we know from X-ray photographs of crystals. Yet they have some sort of motion, oscillatory, presumably, like a pendulum, or the prongs of a tuning fork, for a crystal has temperature just as well as a liquid or a gas, and temperature is a manifestation of the velocities of molecules with reference to each other.

By virtue of this vibration, surface molecules are shaken off more or less rapidly, as determined by the temperature and the kind of substance involved. Many crystals, such as gems of every sort, do not evaporate at all at ordinary temperatures, while others, such as the snow crystal and the crystals of carbon dioxide evaporate freely.

The sources, then, of the water vapour in the atmosphere are sublimation from every snow field and ice sheet; evaporation from water surfaces, soil and vegetation; and volcanic steaming from within the earth—negligible in comparison to the least of the others.

HUMIDITY

Humidity is just another word for moisture or dampness; but "the humidity" means the degree of wetness of the air, expressed either as "absolute humidity," that is, the quantity of water vapour present per unit volume, or "relative humidity," the ratio of the amount of vapour actually present in any given volume to the amount necessary to saturate that volume at the same temperature. The mass or quantity of vapour necessary to saturate a given volume increases rapidly with increase of temperature, but is not materially affected by the presence or absence of other gases. The methods of measuring the humidity of the air, a weather or meteorological element of great importance.

Relation of Humidity to Height

Although water vapour gets into, or becomes a part of, the atmosphere at the surface of the earth, it does not stay at that level, but spreads out by diffusion, is mixed with the other gases of the air by convection and turbulence, and drifted about by every wind from the lightest to the strongest. Of course, if the atmosphere consisted of water vapour and nothing else the density of that vapour would be greatest at the surface of the earth (except wherever, as rarely might happen, the temperature decreased very rapidly with increase of height) because the pressure on it would be greater there than at any higher level by the weight of the vapour up to that level.

But as water vapour is only a decidedly minor member in the atmospheric gas association it has but little control of its own fortune, and so is made to behave in ways that often are not its own. One of the most important of these is its distribution with height. It is not always, nor nearly always, densest at the surface, nor does its density decrease with height according to that beautiful law and order that normally applies to the atmosphere as a whole and that would apply, as stated, to the water vapour if it existed entirely alone.

In the early morning, for example, after a warm calm night, especially when the ground is wet, or where considerable bodies of water are close by, the absolute humidity, and often the relative, too, is greatest close to the surface, and decreases with height for 50 to 100 feet or more. After sunrise, however, vertical convection begins, especially when the sky is clear, and frequently an appreciable wind also sets in, and each of these tends to so redistribute the vapour as to make it the same fraction or percentage of the total air at every level from top to bottom of the agitated or mixed up layer, perhaps (and often) 1,000 feet deep.

This evidently decreases both the absolute and the relative humidity near the surface and increases each of them in the upper portion, owing to decrease of temperature, to and beyond the saturation point, and thus causes the formation of cloud. When cloud is produced there obviously is left just that much less moisture that can be carried to still greater levels, and so it happens that as the temperature decreases with increase of altitude so too does the amount of water vapour necessary to give saturation.

However, the actual amount of vapour over a given place at a particular time seldom if ever decreases nearly so regularly with gain of height. A column of saturated air, for instance, may rise to a considerable altitude and there spread out, forming a cloud layer. This layer in turn may evaporate and the humid but now clear stratum drift on with air of less humidity in either sense, relative or absolute, both below and above it. Furthermore, since layers of air of various origins are thus formed at different levels it follows that, in respect to humidity, the atmosphere at any particular time is quite certain to be more or less stratified, no matter how smooth the curve may be that expresses the average relation of the quantity of water vapour to height.

As just implied, we should expect the curve of the average vertical distribution of absolute humidity to be smooth, that is, to show no tendency towards a maximum or minimum at any height, or, in other words, to show that water vapour does not tend either to accumulate at, or to avoid, any particular level. But this natural assumption is in error. There is a tendency for vapour to accumulate at about the limit of daily convection, the level at which rising air spreads out horizontally. On the average this is about a mile and three quarters high during summer in middle latitudes, and much less than one mile in winter, say six to seven tenths of a mile.

Saturation often occurs also at and near the top of the troposphere, as we know from the frequent occurrence there of the cirrus cloud. The amount of the water vapour per unit volume at the very low temperatures, -60° to -70° F., that prevail at this height, 6 to 7 miles above the surface, is very small, even when the air, or space, is fully saturated. That is why the high cirrus clouds are so thin, gauzy and diaphanous in structure and appearance.

No Clouds in the Stratosphere

As soon as the limit of the troposphere is passed there are no more clouds above to diminish the intensity of the sunshine by day or dim the brilliancy of the stars at night—no clouds here to smoor the sun or blink the stars—except very rarely when either the so-called nacreous, or the so-called noctilucent cloud appears, clouds of wholly unknown origin. The habitual absence of clouds from this upper region is, at first, rather surprising.

As stated above, we know from the cirrus clouds that saturation often occurs at the base of the stratophere, and so we should expect the vapour molecules here to wander on up into the stratosphere—to diffuse, as we say, just as odors diffuse from place to place—and thus to bring about saturation at these upper levels also. We then should except clouds to form at these great heights whenever, for any reason, a distinct drop in temperature should occur. Now, drops (and of course rises also) in the temperature of this region do occur over any given place, drops sometimes of 15° F. to 20° F. and yet the expected clouds do not appear. Why not? Because, in the first place, there is, at best, very little vertical convection in the stratosphere to carry up bodily the vapour from below, while diffusion, the only other way by which it can ascend, is a slow process. Second, before diffusion can carry up much vapour, or any of it to a considerable distance, the air of this region is sharply cooled and the excess vapour, if any, precipitated in the form of ice needles or spicules that fall down gravity far more rapidly than the vapour from which they were condensed originally rose. Now, although this cooling at the base of the stratosphere occurs irregularly, it generally does so at intervals of less than a week.

Hence distinct clouds do not form, even occasionally, in the stratosphere, if we ignore the rare nacreous and noctilucent clouds, because it is kept exceedingly dry by the frequent formation of an invisible haze at its base. Contradictorily, then, clouds do not form in the stratosphere because clouds do form in the stratosphere—never well up and perceptible, because continually and at frequent intervals, hence low down and imperceptible.

Daily Run of Humidity

There are several things that affect the humidity of the atmosphere at a particular place. Some, if not all, of the more important of these are, temperature, convection, nature of the immediate surroundings, wind velocity and wind direction. In general, increase of temperature carries with it an increase in the absolute humidity, through increase of evaporation; and a decrease in the relative humidity, owning to the consequent increase in the vapour capacity per unit volume, a capacity that increases rapidly with increase of temperature. At any rate this is the initial effect. After a sufficient time for much more water vapour to be provided by evaporation the relative humidity may come up to, or even exceed, its former value.

The effect of convection on the distribution of humidity may seem obvious since, as applied to the atmosphere, it com- monly is lower air being pushed up and staying up, with a slight but more widely spread descent of the upper air; or a local descent of the upper air and slight ascent of the neighboring lower air; or, ideally, a mere interchange of position between the upper and the lower air, the one coming down and staying down while the other goes up and stays up. If, then, in this latter, or ideal, case the lower air at first is more humid than the upper, presently, after convection sets in, the upper certainly will be more humid, one might suppose, than the lower, because the lower air (the more humid) has become the upper, and the upper air (the less humid) the lower.

This delightfully simple relation seems, as stated, perfectly obvious, but it is invariably true only for a definition of humidity different from either of those already used. For these it may or may not be true. If the *relative* humidity is greater in the lower air than in the upper, then after convectional interchange it will be even more pronouncedly greater in the then upper air owing partly to the increase of its relative humidity, due to cooling by ascent, and partly to the decrease of the relative humidity of the falling air, owing to its gain in temperature.

Clearly, though, the lower air may be so dry, with reference to an upper layer, that it has the smaller relative humidity not only before, but also after, they have interchanged places. Evidently, then, the relative humidity of the lower air may be either greater or less than that of the upper both before and after the convectional turn-over; it also may be less before and greater after, but as explained above, never greater before and less after. Neither does the simple relation, greater before interchange greater after, less before less after, neccessarily hold for absolute humidity (the mass of water vapour per unit volume), since it decreases with ascent and consequent increase of the volume occupied by a given mass of the vapour and similarly increases with descent.

We thus are brought to the alternative of either giving up the simple and seemingly self evident relation mentioned above or of finding some sort of humidity (so defining it) that is not affected by change of temperature or change of volume. Such a definition may, at first, seem impossible, but so long as water is not removed by precipitation these conditions are fully met by "specific humidity," the mass of water vapour per unit mass of the humid air. This particular humidity (as per definition) is not so extensively used as either of the others, relative and absolute, but it is in use, and occasionally, as in the case under discussion, is, by far, the most convenient.

So much, then, for the effect of convection on the humidity of the air—not exactly what one might have supposed, but for that reason all the more interesting. And there are other things that affect humidity. It was stated, a page or two back, that it varies also with the nature of the immediate surroundings. The extreme cases of a small island in midocean and a sand dune in a desert will make this effect of environment practically self evident. On the desert there usually is very little water available for evaporation, so that the air once dry, for whatever reason, normally remains dry for a comparatively long time.

In the case of a small island, however, the situation is radically different. Here there is water in every direction that always keeps the lower air, at least, quite humid.

It should be noted in passing, however, that while high humidity is essential to precipitation, that result does not necessarily follow. Catalina Island, for example, is as arid as any portion of the adjacent mainland. High humidity is a factor essential to precipitation but it is not of itself alone entirely adequate. There must coexist also an active process, such as vertical convection, that tends to push the humidity to supersaturation, and would so push it if condensation did not occur.

Another of the modifiers of humidity is wind velocity, and it acts in different ways. It expedites evaporation—the wind, as we know from abundant experience, rapidly dries up wet roads, dries them by increasing the rate of evaporation of the water on them. This increase of evaporation gets moisture into the air at a correspondingly increased rate and thereby affects the humidity. Furthermore, winds are accompanied by turbulence, a mixing of the surface air with all the air for some distance above. In this way the humid air over damp soil and bodies of water is mixed with the commonly drier air overhead, and thus the humidity at all levels in the agitated layer more or less altered.

Also wind direction is an important factor in determining the humidity of nearly all places except, perhaps, those of middeserts and tropical oceans, where its effect is at a minimum and even uncertain as to sign. Evidently the air is more humid when it has just come from over a warm body of water than it generally is when from a dry land area. The sea breeze therefore is more humid, as a rule, than the land breeze, and the summer monsoon winds richer in water vapour than the counter winds of winter.

From the foregoing controls of humidity it is evident that at low levels over the ocean it is quite likely to increase during the day, especially when the wind is light, owing to the addition there by evaporation of abundant fresh vapour, and the slight extent to which it is carried off by convection and mingled with the upper air. Obviously, though, this can go on indefinitely, as on the average it does, only because the accumulating vapour is more or less regularly removed from the place in question. This necessary removal is effected locally and temporarily by the horizontal drift of winds and vertical mixing by turbulence, and eventually completed somewhere else by condensation and precipitation.

Over the ocean, too, as there is very little change of temperature between day and night, and but feeble convection, there are no marked humidity maxima and minima at approximately uniform hours, as there are over most land areas. On land, as just implied, the humidity, both relative and absolute, commonly undergoes distinct diurnal fluctuations in the warmer season, not in the colder, with maximum relative humidity of the lower air at the time of minimum temperature and minimum at the time of maximum temperature; and maximum absolute humidity about 8 to 9 o'clock a.m. and again at 9 to 10 o'clock p.m. with, of course more or less pronounced minima at some intervening time, both day and night, as determined largely by wind and convection.

During winter the diurnal changes of humidity are comparatively slight since evaporation then is slow, the water being cold, and diffusion incident to thermal convection practically absent. Since change of temperature so greatly affects humidity, and since temperature depends largely on sunshine, it follows that the daily course of humidity is greatly affected by the amount of cloudiness—the greater the cloudiness the less the sunshine, and the less the sunshine the lower the temperature and the less the evaporation.

The Yearly Run of Humidity

Since water vapour is gotten into the air by evaporation, a process that varies directly with the temperature of the evaporating water, and distributed or diffused by horizontal winds and vertical convection, it follows that the annual variation of humidity is very much like unto its diurnal variation. Hence, throughout the first mile or so above the surface of the earth, the absolute humidity of the air is greatest, as a rule, and its relative humidity least, during the summer; and the relative humidity greatest, and absolute least, in winter. From this in turn it follows that clouds commonly are lower in winter than in summer, both over land and over sea. Of course, too, the annual run of humidity, like the daily, also depends on locality. Over tropical seas there is but little variation, while in certain other localities the change is all the way from frequent saturation at one season to distinct aridity at another.

Horizontal, or World, Distribution of Humidity

From the fact that all sorts of humidity, the familiar absolute, the more familiar relative and the specific, vary so greatly with temperature, nature of the surroundings and the direction and speed of the wind, it follows that each kind varies also with latitude and longitude as well as with the time of day and from season to season. However, it is not necessary here to consider specific humidity, the weight of water vapour per unit weight of humid air, since it generally varies in the same manner as the absolute humidity, or weight of water vapour per unit volume. The variation of this latter, or absolute humidity, with location is amazingly great. On the most stiflingly hot and humid days in a tropical jungle, for instance, the humidity may amount to even 5 per cent of the air by volume, or 3 per cent, roughly, by weight, while in the coldest places of the world when the temperature is lowest it may be even less by weight than one part in 300,000.

Thus the greatest amount of water vapour that sometimes is in the air where the temperature is quite high is at least 10,000 times greater than that which is or can be in it at temperatures occasionally reached in eastern Siberia. Usually, too, the absolute humidity is quite small over all desert regions, but probably never nearly so small as it necessarily is wherever and whenever extremely low temperatures prevail. In general, and owing to the very rapid rate at which the amount of water vapour necessary to produce saturation of a given volume increases with temperature, absolute humidity grows rapidly and continuously with decrease of latitude from either pole to the equator, except across the high pressure belts of 25° to 35°, roughly, both north and south—horse latitudes of the oceans, and desert zones of the continents.

Quite commonly the absolute humidity is less along these belts than on either side of them, just as they are also zones of minimum cloudiness and rainfall. The dryness of these zones, in terms of amount of water vapour lacking per unit volume to produce saturation, is more pronounced over the land or desert areas, where there is but little surface and plant water available for evaporation, than along the horse latitudes of the oceans. But strange as it may seem, even these sea areas are arid regions in the sense of having relatively low humidity and scanty precipitation. This is because the air here usually is slowly descending and for that reason commonly rather dry except near the surface.

In addition to these world-wide variations of absolute humidity with location there also are many others of lesser extent. For instance, it decreases in general with distance from the ocean, but not in any regular sort of way. Where the rainbearing winds are nearly all from the same general direction, as they are, for example, over the central United States, they carry, on the average, less and less water vapour, hence exhibit a progressive decrease in absolute humidity, with increase of path over land after leaving the ocean. This is owing to the fact that the amount of water vapour taken up into the air by reevaporation over any particular area is less than the amount removed from it by rainfall on that same area—less by the contribution of this area to the stream flow, or runoff.

But the winds from the different oceans do not all carry equal amounts of vapour, nor do they, as a rule, penetrate equally far into the land, hence the place of lowest average absolute humidity generally is considerably removed from the most inland point. Furthermore, the decrease of absolute humidity with progress of the wind from the ocean often is very irregular. Thus a coastal range of mountains, where the prevailing winds are on-shore, commonly has considerable to heavy precipitation on its windward slopes, while to the leeward the humidity usually be so low as to produce semiaridity for hundreds of miles beyond, unless another mountain range is encountered. And even if such mountains do lie across the winds, and precipitation in quantity does occur on them, the average absolute humidity is not thereby increased, but more or less decreased in proportion to the contribution of the given region to the general stream flow.

The most important things to remember in considering what the distribution of absolute humidity is, and why, are the facts that its possible value, and generally, therefore, its actual value, except in arid regions, rapidly decreases with decrease of temperature; that it is decreased by precipitation; and that it is increased by evaporation. Change of pressure, and change of temperature, by altering the volume occupied by a given mass of air, also change the current value of the absolute humidity.

These same things, mainly, evaporation, precipitation, change of pressure, and change of temperature, also are the factors that control the relative humidity, that is, the current fraction of saturation, or ratio of the amount of water vapour actually present per given volume to the amount necessary to produce saturation at the same temperature of the same volume. It should be especially noted that while the actual temperature is essential to the determination of at least the maximum possible value of the absolute humidity, it is of no importance in fixing the value of the relative humidity.

Since the vapour capacity of a given space or volume—the maximum amount of water vapour that can exist in that space in the presence, to be exact, of a flat surface of pure water-depends on temperature only and rapidly increases with increase of temperature, it is evident that, with a given amount of water vapour, short of that necessary to produce saturation, present per unit volume, any decrease of temperature will increase the relative humidity. This is owing partly to decrease of the volume, the pressure remaining the same, occupied by a given quantity of the vapour and partly to the decrease of the vapour capacity per unit volume, both incident to the decrease of temperature. Conversely an increase of the temperature, other things remaining equal, decreases the relative humidity, and so too does decrease of pressure, temperature remaining constant, owing to the corresponding increase of volume.

From these fundamental principles it is practically obvious how the relative humidity must vary with change of place. Thus, tropical winds on their course to higher latitudes normally become colder, and therefore gain in relative humidity even to saturation and the production of fog and cloud.

Similarly the relative humidity of winds from higher latitudes generally decreases, owing to increase of temperature, except where evaporation is particularly rapid. Again, in the summer, relative humidity commonly decreases with progress inland over a continent, owing partly to increase of temperature and partly to an actual decrease of the humidity in the air. In the winter the relative humidity generally is greater over continents than over oceans, owing to the comparatively low temperature then of the land surface. Naturally, too, the relative humidity is especially low over arid regions, and this condition extends in a measure to the oceans, where the relative humidity averages about 75% along the horse latitudes, and 80% over the rest of the ocean areas.

CONDENSATION

Condensation, or the changing of water from the gaseous state to the liquid state, is exactly the reverse of evaporation. And just as the latter, that is, evaporation, makes the space into which the vapour passes humid—makes the air humid (part water) we commonly say—so obviously its reverse, condensation, always reduces the absolute humidity, or amount of water vapour present per unit volume. These two processes, evaporation and condensation, are the fundamental controls of humidity, every form of which, so long as the temperature and the pressure are kept constant, is increased by the one and decreased by the other. It, therefore, is desirable to know something of the conditions under which this important phenomenon, condensation, occurs, and to formulate some idea as to how it occurs.

It is now well established by evidence fascinating and convincing, but too long to include here, that each pure gas, hydrogen, oxygen, water vapour or what not, consists of equal discrete particles so infinitely small that if 500,000 people were each picking them out at the rate of 100 per minute, day and night, it would take them 1000 years to accumulate particles enough to make a volume of the air we breathe as large as the head of a pin. And yet we know to closer than one part in a hundred the weight or mass of each of these particles, and how any kind, oxygen for instance, differs from any and every other kind. We know, too, that all these particles are in swift motion, on the average, among themselves with, seemingly, all the irregularities of a swarm of bees, and more besides. Yet in spite of this apparent utter confusion as to direction, and wide range of velocity, certain surprising regularities or laws do obtain.

Thus the average velocity of the molecules varies directly as the square root of the absolute temperature, inversely, as the square root of its mass. In a mixture of gases, such as constitute the atmosphere, the average energy of each molecule is the same as that of any other—whatever the lighter ones lack in mass they make up for in speed. And that speed is astonishingly great, in fact at ordinary temperatures it is around 1500 feet per second, or say that of an ordinary rifle bullet, while the velocity of the water molecule in the is approximately 1900 feet per second.

But regardless of their great velocities the molecules of a gas, or anyhow much the greater portion of them, seem to come to rest when they impinge onto a solid or liquid surface. That is, at such times they at least momentarily condense where they strike. If the temperature of the object struck is high they are quickly knocked off again, but if its temperature is low, if, in other words, its molecules vibrate slowly, the chance that the arrested molecules will promptly leave it is correspondingly less.

Hence, at a certain temperature which we call the dew-point more moisture collects on a water surface, and also on many other objects, than leaves it. This is condensation. When this condensation occurs on foreign particles in the atmosphere, however minute they may be, the result is a fog or cloud. When every thing of this dust-like nature is taken out of the air, by filtering it through a tube stuffed with raw cotton, or otherwise, fog does not form immediately the dew-point is passed, nor indeed under any degree of supersaturation that possibly can occur naturally and in the open. We know, therefore, that if the atmosphere were pure air, in the sense of containing nothing that could be mechanically filtered out of it, never a fog nor cloud would form, nor rain nor snow, only evaporation from the surface, and condensation again somewhere on the surface—evaporation chiefly in the warmer regions and condensation in the colder.

The atmosphere, however, is not pure, but filled with countless myriads of condensation nuclei of many kinds—kernels as it were about which water shells can and do form, immediately the state of saturation is passed. Though but an infinitesimal part of the ordinary outdoor air they nevertheless are absolutely essential, indeed weather vitamins without which there could be no rain and no snow.

Clouds Yield Rain

The above may describe the making of a cloud droplet, but a cloud droplet is a long way from being a rain drop. Indeed it takes about a million of the former to make one of the latter, and besides the droplets have but little tendency to spontaneously and rapidly unite, as is evident from the fact that clouds often float along for hours and even days without yielding a drop of rain. On the other hand, rain often does occur and so we wonder why clouds sometimes produce abundant rain and at other times none at all.

They look quite alike on the two occasions, as like as peas in a pod, and are alike except that when rain is falling a lot of relatively large drops are present while none are present when no rain is falling. But that is the point. Why are the clouds not always the same, always consisting entirely of minute droplets or always containing also rain drops? Well, as stated, there is no important difference between these clouds except in respect to the detail of the size of some of the drops, but there is a marked difference in the movement of the air in which they occur, and in this rests the secret we are trying to find. Clouds from which no precipitation, rain or snow, is falling, are floating along in air that has very little vertical movement, whereas there always is appreciable, and at times even great, velocity upward of the air in which rain drops or snowflakes are forming.

Now, water drops fall through still air with a velocity that varies with their size, from about one tenth of an inch per second in the case of the cloud particles, to some 25 feet per second for the largest possible rain drop—one fifth of an inch in diameter, approximately. Clearly, then, whenever the movement of the air has an upward component, clouds will form as soon as expansion has brought the temperature to the dew-point, and will so continue to form in the ever fresh air at the same level as long as that particular stream continues to rise. Yet, despite this constant formation of cloud not a droplet can fall until by growth from condensation, or coalescence, or both together, it has become large enough to overcome by its weight the upward drag of the rising air.

And, further, since more and more cloud is formed, and since well up in the cloud the velocity of ascent grows less and less with increase of height, and finally ceases altogether, it follows that after a time there must be a falling of drops of appreciable size; in short, a rain, and at such rate as to equal the fresh supply of cloud, or new condensation.

Other Forms of Precipitation

Rain is not the only form of precipitation: *Snow.*—When the condensation occurs at temperatures below the freezing point the usual result is a snow crystal, and when, in the course of their fall, many such crystals get entangled together into a single mass the result is a snowflake. *Sleet.*—Sometimes rain falls into a freezingly cold layer of air only a thousand or so feet thick and reaches the earth as frozen pellets, or sleet of the variety that rattles when it hits a windowpane.

- *Hail*: On the other hand, in a vigorous thunderstorm rain drops frequently are carried to such great heights that they freeze, and often grow to considerable size before they overcome the lift of the rising air and fall to the ground as hail stones.
- *Graupel*: Still another form of precipitation is the rather unfamiliar graupel or soft hail—white, granular bits of ice roughly spherical in shape and an eighth of an inch, more or less, in diameter. It may be formed in subcooled clouds, that is, clouds of water droplets at temperatures below the freezing point, much as rime, or granular snow tufts, are formed on the windward side of exposed objects in a drifting, subcooled fog. Certain particles here and there in the cloud may, and undoubtedly do, fall faster than most others, and grow by capture as they fall. If, now, the cloud is subcooled the droplets are certain to freeze immediately they come into contact with the falling larger mass and adhere to its underside, thus giving it, as it turns every·which way in the course of its fall, an approximately spherical form.
- *Fog drips*: The drippage from trees and other objects when a dense fog drifts through them. *Dew* and *frost* are intentionally omitted from the above list of precipitation form, because they do not fall from the clouds as do rain and snow.
- *Quantity of precipitation:* What is the quantity of all this precipitation? Well, for the whole world there is lots of it. There is enough humidity in the atmosphere all the time to make a layer of water at least an inch deep over the entire earth, while the average rate of the total precipitation, and of course also the rate of evaporation, is approximately 16,000,000 tons per second, enough to cover the whole earth in the course of a year with a layer of water three feet deep.

Atmosphere usually is not Saturated

Evidently this precipitation furnishes all the water used in plant growth, and all that runs off as stream-flow except, to be exact and if geologists are right, a portion of that negligibly small amount that comes from gysers and certain hot springs—a portion furnished, some hold, by the primordial rocks themselves. But there is another important thing that results from the occurrence of precipitation that is not so obvious. It is the fact that in this way the atmosphere is kept from being everywhere and at all times intolerably humid, indeed saturated, or very nearly so. Every rain and every snowfall takes that much water out of the air—dries it to that considerable extent and thus keeps the humidity generally far below the saturation point, and therefore normally within the bounds, not of endurance only, but of actual comfort.

DISTRIBUTION AND CHANGES OF ATMOSPHERIC PRESSURE

Although atmospheric pressure, or weight of the air per unit horizontal area directly above the place where the pressure is measured, is not a weather element of any particular importance so far as our feelings are concerned, such as temperature and humidity are, nevertheless it is the element that more than any other locates the weather of today and tells us what tomorrow's will be, and therefore is of great importance. Latterly this importance has appreciably decreased, however, because other methods have come into prominence for finding the positions, magnitudes, speeds and directions of travel of the contrasting masses of air whose conflicts and whose passage keep the weather in ceaseless ebb and flow from fair to foul and foul to fair.

Still these other methods have not yet been sufficiently developed to justify ignoring the distribution of atmospheric pressure, and how it is increasing here and decreasing yonder, as aids to a foreknowledge of the weather of the morrow. Neither is it at all likely that so great a development and use of these other means, while entirely possible, will be effected at an early date, if indeed they ever are so developed. Therefore a knowledge of the distribution of atmospheric pressure, the near equivalent of the distribution of the air itself, and of its changes, still are, and for a long while must remain, matters both of great practical importance and of keen theoretical interest.

Barometer Corrections

Since the pressure of the air at any particular time and place is, as stated, the weight of all the atmosphere above a horizontal unit area then and there it follows that a knowledge of the values of this pressure over a given region, however extensive, tells us, when the proper corrections have been made, just what the distribution of the air itself is over that same region. But what are these correc- tions? That depends on the kind of instrument used to measure the pressure and on what we want to find, pressure or quantity of air, for we must not forget that the same quantity or mass of air, as also of other things, varies in weight, or pressure, with position—not with every change of position but with most changes.

But we can think easier and straighter if we consider one thing at a time, and know exactly what that one thing is. Let us first, then, compare the *masses* of air over different places. We can do this by means of a mercurial barometer which is only a device for balancing the pressure exerted by a column of mercury at its bottom, tending to make it flow out, against the pressure there of the atmosphere that keeps it back in.

Clearly the pressure exerted by the mercury at the bottom of the column, when corrected for the capillary action of the tube, is proportional to the product of its density by the height of the column by the force of gravity, or pull oil a unit mass by the earth, at that place. Similarly the pressure there of the atmosphere is proportional to the weight, or mass times the force of gravity, of all the air, if still, above that level (the level of the bottom of the mercury column) per horizontal square inch, or other specified area.

When the air is in motion, however, as it always is, a further correction of the reading of the barometer is indicated, since the weight (downward push) of a given mass varies with the direction and speed of its motion.

But since the value of this correction (never known accurately) at most is negligibly small it will not be discussed here —just mentioned for the sake of completeness.

Similarly the effect of the variation of gravity with height also is negligible. In this case then we may consider sauce for the goose as sauce for the gander too, in the sense that gravity applies in equal (almost) measure and exactly alike for both the mercury and the air. Hence the mass of the air above any given horizontal area, such as a square foot, say, at the level of the mercury in the barometer basin, is proportional to the height of the mercury column, or reading, as we say, of the barometer, corrected for whatever slight difference in that height there may be due to the pull, or capillary action, between the mercury and the wall (glass) of the barometer tube, and corrected also for the density of the mercury—corrected to a common or standard density, commonly that which it has at the melting point of ice, 32° Fahrenheit, or 0° Centigrade.

In this way the readings of all the mercurial barometers are made strictly intercomparable, and the ratios of their several readings to each other exactly the same (to within an unknown but always negligible amount) as that of the masses of air above them, so that a geographic map on which these values, if simultaneous, are written, or otherwise represented, is also a map of the distribution of the atmosphere over that same region at the time the readings were made. To obtain the instantaneous distribution of the air the barometer readings necessarily must all be taken at that particular time, whatever it is, 8 A.M. 75th meridian time, or what not.

If, however, aneroid barometers are used instead of mercurial for obtaining the mass distribution of the atmosphere, then their readings, which indicate pressures directly, must be corrected for the local values of gravity, because the atmospheric pressure, being the weight per unit horizontal area of a vertical column of air, varies directly with the force of gravity which, in turn, increases with latitude. This quantity, or mass, distribution of the atmosphere is of considerable importance, especially through its influence on the rate of transmission, in or out, of radiation. A thin atmosphere and a thick one are to the earth much like a sheet and a blanket, respectively, to one sleeping in the open.

Although this mass distribution of the atmosphere is very important as a factor that helps to make the weather, especially as to temperature, what it is, still very little use ever is made of it. The distribution of pressure, or push, in the air is on the contrary frequently, carefully and extensively plotted because it is the unequal distribution of push on the atmosphere (or pressure, for they are equal to each other) that makes the winds to blow and thereby, in large measure, but not wholly, change the weather of one place by bringing to it the air and the weather of another. Clearly, then, a knowledge of this distribution must be, and is, a great help to the fore- caster when judging of the weather of the morrow and the day after. That is why the barometer is so extensively used in making meteorological observations and used almost exclusively for measuring the atmospheric pressure.

Now the pressure exerted by a column of mercury, the column in this case that is balanced by the pressure of the air, is proportional to the height of the column (corrected for capillary drag), to the density of the mercury (which in turn varies with the temperature) and to the local force of gravity. Hence, to compare the *pressures* at different places, one with another, it is necessary to correct the readings of the barometers to what they would be if all the instruments had the same temperature, say the temperature of melting ice, and were actuated by the same force of gravity—its value at sea level at latitude 45° N., or some other agreed-upon value.

The readings of aneroid barometers require no corrections for this purpose since, as previously stated, these instruments measure pressure directly.

To sum up: If we are measuring the mass distribution of the air the readings of the mercurial barometer must not be altered for gravity differences, while those of the aneroid must be so changed. On the other hand, when it is the distribution of pressure we are finding, the element we nearly always seek, it is the readings of the mercurial barometer that must be corrected to what they would be under some common force of gravity and not those of the aneroid barometer—they are true pressure values, if the instruments are correct, as they stand, and therefore must not be changed.

The next and final step is to reduce all these pressure values, or corrected readings of the barometer, to what they would be if taken at some common level. This is done because the travel or flow of the air, due to difference of pressure, is, of course, along the route of least resistance, which route, as a rule, is over a nearly level surface, that is, a surface everywhere normal to the direction of the force of gravity, such as the mean surface of the ocean. Actually the direction of least resistance to a sample of air through the surrounding atmosphere, the direction indeed of no resistance at all, save alone that incident to viscosity, is along what the physicist calls an isentropic surface. However, this exact and refined treatment of the subject will not here be followed to its conclusion, though to do so affords both surprises and mental satisfaction.

But let us have as little boggy ground as possible under our mental feet. Why is the level surface usually the one of least resistance to a sample of moving air? Because ordinarily the change of temperature of the atmosphere with change of height is such that work would have to be done on the sample to carry it to a greater height or push it to a lower level, but no work (save that due to viscosity which is the same in all directions) to move it horizontally.

Normally this distribution of temperature is such that a sample of air taken upward would cool, incident to its expansion under a decreased pressure, to a lower temperature than that of the then adjacent air. Being at the same pressure as this adjacent air but cooler, it also would be denser, so that an actual lift would be required to carry it above its initial level. On the other hand, if taken downwards the compression due to the increased mass of air above it would warm it to a higher temperature than that of the other air at its new position. But being now warmer than the adjacent air, and having the same pressure, it also must be lighter. Hence, normally, only an actual downward push can force a sample of air at any height to a lower plane. In short, air normally can be taken out of its own level only at the cost of work, work against gravity if pushed up, and work against the buoyance of the denser air beneath if forced down. Clearly then no work against gravity, nor any against buoyancy, is required when, and only when, air is moved over a level surface, or, to be exact, over an isentropic surface—a surface which, as already explained, usually is very nearly level. Obviously any direction over this surface is a direction of least resistance.

Therefore if we would know the magnitude of the cause of the winds, and from that magnitude infer their strength, we must find the values of the atmospheric pressure at the *same* level at various places over the region in question. If we had a great number of these pressure readings somewhat evenly distributed over an area of considerable extent, say the size of the United States, or larger, we then could draw on a map of that region a line connecting all the places that, at a particular instant, had the same pressure. This line, according to present practice, we would call all "isobar," or line of equal pressure. By drawing a number of such isobars, corresponding to small but constant differences in pressure, we construct a map or diagram from which the horizontal pressure gradient, or horizontal push per mile, say, can be found. Note that no two isobars (lines of equal pressure) can cross each other. Such crossing would signify two different pressures at the same time and place, a manifest impossibility.

The direction of the flow of the air under these unequal pressures alone would be horizontal, as previously explained, and along the course of the shortest distance between the lines of equal pressure, that is, normal to these lines. A sample of air would not flow off slopingly to this normal to the isobars, though that too would be going from a greater to a lesser pressure, because every component of pressure directed to the right along or parallel to an isobar is exactly balanced by a component of pressure in the opposite direction. The only unbalance is strictly normal to the direction of the isobars, and therefore this is the direction of flow, or would be, if there were nothing to disturb or prevent such flow. As a matter of fact this flow is modified by the rotation of the earth, and so profoundly modified as to change it from the direction of the push to the right angles thereto, except in the lower air where the drag of the surface makes the change in direction less than a right angle. But this fascinating paradox must be passed by. Its explanation would be out of place here; besides only the language of mathematics properly fits it anyway.

Since the movement of the air under differences of pressure is horizontal, or very nearly so, as just explained, it is clear that our isobars must be drawn for a common level if we would know from them what the values of the pressure gradients are that are making the winds of the moment to blow, fixing the intensity of a passing storm and largely determining whither and how fast it shall go. But what should be this common level, and how can we find the value of the pressure there? These questions are old; their answers would be new.

For the mariner they are simple enough: sea level is the place, and read our barometer where we are (on that level) is the way. But for the landlubber it is a different and awkward matter, since in general no two of his observing stations are at the same level. For the purpose of weather forecasting one can chart on his map the departures of the barometer readings of the several stations from their respective normal values for the season, or day, in question, and then connect by a smoothly curved line the places of equal, like departures. Such a map can be, and has been, used with fair success, but it never became a general favorite, though it is the only pressure map that deals with known accurate magnitudes.

One thing that renders such a map less helpful to the forecaster than at first one might expect it to be is the fact that the normal, or average, pressure for the given month, say, is one thing for weather of one kind and another for weather of a different kind. In short, the general average or normal for all days is not applicable alike to all sorts of weather conditions—the normal itself, or the observed departure from it, needs to be corrected for each special reading, but by how much we do not know. The pressure-departure map therefore is not entirely satisfactory and can not be made so.

Another promising procedure, and one that has been seriously attempted, is to use the pressures that exist at some definite height above sea level, half a mile, one mile, or what not. Wherever this level is above the surface, surely at that level there is air. Surely, too, instruments can be carried to that level and the actual pressure there measured with considerable accuracy.

However, this direct manner of obtaining the desired pressure values is now too expensive to be practicable. Therefore the method used has been to compute the pressure at a given height from the pressure and temperature at the surface with a general formula and then correct the value found according to the circumstances, especially as to season and direction of the wind, as empirically deduced from data obtained under various conditions by kite, balloon and airplane. But at best more or less doubt as to accuracy always attaches to the values thus obtained. Hence weather maps that ostensibly show the distribution of atmospheric pressure over a horizontal surface at some definite height above sea level have not yet come into general use.

There remains, in this connection, one other obviously possible procedure, the one that has been, and now is, all but universally and solely used. This is, to reduce each pressure reading made on land to what it presumably would be at sea level immediately beneath the point of observation if all below that point, and round about, were not rock and soil, but air in that state and condition appropriate to the location, season and kind and distribution of the weather. The calculation, only approximate of course, is based on the known height of the station above sea level, and the observed temperature and pressure of the air, and then empirically more or less corrected. The chief source of error is the local surface temperature.

Sometimes this is abnormally high and sometimes abnormally low. In the first case the reduction to sea level is made on the assumption that the hypothetical air beneath the place of observation is correspondingly warm and light, and in the second that it is correspondingly cold and heavy. The one yields a pressure too small and the other a pressure too great. Occasionally this error is so big that even the sense of the pressure distribution is reversed—actual cyclones and anticyclones, as indicated by the direction of the winds, changing by reduction to anticyclones and cyclones, respectively, on the map. Pity then the forecaster. But don't pity him overly much. He doesn't deserve it. He just winks the other eye, having in mind any poor devil trying to learn the game from such charts, mentally alters the map in accordance with his background of experience and all data available, some of which are never charted, and goes ahead serenely, if not always confidently. At his own game, and with loaded dice, of course he beats the innocent novice—until he, too, gets wise.

The one thing we definitely know about all reductions of the barometer to sea level, or any other level, is this: They invariably are in error. Generally, however, they are usable, and we do not know of any practical way to improve them. We know too that a "level" surface is not the correct surface on which to draw isobars; but then the surface that is the proper one, the isentropic surface, does not "stay put," as we say, but perpetually so warps and waves as to make pressure reduction to it quite out of the question. As a rule, however, this correct surface is, as previously stated, nearly level. Hence, and because reduction of pressure to a common height usually is only slightly erroneous, it follows that the sea-level map meets the needs of the weather forecaster fairly well, and better indeed than any other that as yet it has been practical to construct.

In this connection let us confess a little inconsistency and confusion of practice in our sea-level reductions. We reduce the pressure, or barometer, readings to sea level, but we leave the temperatures unchanged. Hence our weather maps show sea-level pressures and "mile-high" temperatures (themselves often exaggerated by surface conditions) side by side. This for the forecaster. When the climatologist sums the weather data all up he commonly reduces both temperature and pressure to sea level. Consistency, thou mayest be a jewel, but art not always prized.

Chapter 4

Atmospheric Pressure

Since the atmosphere is so exceedingly fluid, moving horizontally under the slightest push, one might suppose that its sea-level pressure would everywhere be the same and therefore its mass distribution substantially equal the world over. However, no such uniformity obtains. In fact, this pressure varies from place to place at any time and from time to time at any place, seasonally, diurnally, semidiurnally, and irregularly in different manners.

And there are two fundamental causes of all the larger and more important of these changes of pressure; namely, differences in temperature and the rotation of the earth. Because, on the average, the atmosphere for miles above the surface of the earth (not all the way out, however) is warmer in equatorial regions than at middle to high latitudes, it there is most expanded and from there has most overflown towards either pole. Hence for this reason alone, if there were no other, the sea-level pressure of the air would be continuously lower over the tropical portion of the earth than over the colder parts on either side.

But there also is another important factor contributory to the final result, namely, the rotation of the earth, in consequence of which the circulation, just mentioned, of the air between the equator and the poles seldom moves with the same speed and in the same direction as the surface of the earth itself, though this is the state, a dead calm, towards which surface friction and turbulence always and everywhere tend to bring the air. The normal resultant of this tendency, rarely anywhere satisfied, is a system of more or less continuous winds from easterly points between the latitudes 30° N. and 30° S., roughly, and a system of westerly winds in each hemisphere between the latitudes 30° and 60°, or thereabouts. The winds beyond latitude 60°, especially those of the southern hemisphere, again are more or less from the east.

But what has this to do with the pressure of the atmosphere? Much. The bulge of the earth is just such that a frictionless object on its sea-level surface if once at rest would remain at rest in perfect equilibrium, and such would have been the case if the rotation of the earth had been either greater or less than it now is. In the first case the bulge would have been correspondingly more pronounced and in the second less pronounced. If therefore our frictionless object were given a velocity eastward it would find the earth insufficiently bulged, so that to it equatorwards would be down hill. Similarly, if it were given a velocity westward its down hill then would be in the direction of the adjacent pole. Hence the winds of the tropical regions, being from the east, tend to drift off polewards, those south of the equator towards the south pole and those north of it towards the north pole. Therefore in this mechanical way, as well as by virtue of the expansion and consequent overflow of the relatively warm air, a belt of low sea-level pressure is established and maintained along, and for some distance on either side of, the equator.

Also, the poleward push of the tropical winds is roughly balanced by the equatorward urge of the mid-latitude winds- winds prevailingly from the west—from which there results two belts of high pressure, one in the northern hemisphere with its axis roughly along latitude 30° N., and the other similarly situated in the southern hemisphere. Beyond each of these belts of high pressure in turn, and for exactly similar reasons, there is a belt of low pressure with its axis at about latitude 60°. The one in the southern hemisphere is by far the better developed of the two. All along it the average pressure is decidedly lower than it is anywhere else on the globe. The current pressure there, however, undergoes frequent and great changes owing to the passage of storms as presently to be explained. Beyond these circumpolar belts the sea-level pressure again increases, as is evident from the fact that there is a more or less continuous outflow of surface air from each of the polar regions.

In respect, then, to the major features of the distribution of the atmosphere over the earth there are seven distinct great regions: three where it exists in smaller amount than the average for the world as a whole, namely, the equatorial belt and the two 60° latitude circumpolar belts; and four, the two high pressure belts of latitudes 30° N. and 30° S., and the two polar caps, where it occurs in amounts larger than the average.

Centers of Action

Not only are there seven permanent divisions, as just listed, in the distribution of the atmosphere and its sea-level pressure over the earth, but three of these divisions in turn have important semipermanent subdivisions, which, owing to their connection with the direction and strength of winds, and course, intensity and frequency of cyclonic storms, commonly are called centers of action. Several of these are the peaks of high pressure that normally exist near the eastern ends of the ocean portions or segments of the high pressure belts. These belts are far better developed, and undergo fewer and smaller changes over the oceans than on land, but even here they are not of equal intensity from end to end. Near the eastern limit of each of these links the pressure normally is much greater than it is elsewhere, a circumstance quantitatively related to the strength of the surrounding winds. As this pressure maximum shifts in position and intensity so do the winds of that general region also vary in place and strength.

Evidently there must be, and there are, five of these highpressure centers of action, one each in the North Pacific, South Pacific, North Atlantic, South Atlantic and Indian Oceans, respectively.

Besides these there are two recognized low-pressure centers of action, namely, the Icelandic low, southwest of Iceland at about latitude 60° N., and the Aleutian low, just south of the eastern Aleutian islands and near latitude 55° N. The former of these, the Icelandic low, though shifting in position, and often being replaced for short intervals by high pressure, or an anticyclone, nevertheless statistically, or on the average, is present month after month and year after year as determined by the distribution of temperature over and round about that region. The Aleutian low, on the other hand, is only a winter phenomenon when the land areas round about are much colder than the water where the prevailing low pressure occurs. In the summer time when the land areas are comparatively warm, low pressure is not a normal feature of the ocean in the general region of the Aleutian Islands or the Gulf of Alaska. Both these "lows" also frequently are called centers of action since many migratory cyclones either start from them, or, at least, come by way of them from regions beyond.

There are no definitely known counterparts to these cyclonic, or low-pressure, centers of action in the southern hemisphere. Perhaps, however, they do exist, one over the Ross Sea where the surface temperature obviously is higher than that of the mostly surrounding great glaciers and snow fields, and possibly another, but much feebler one, over the Weddell Sea.

Pressure Changes

As every one knows, where the air is being heated it expands and where it is being cooled it contracts. Therefore the atmosphere overflows from the summer hemisphere onto the one where winter prevails, and thereby causes a difference between their respective quantities of air and values of their sea-level pressures. The change is small from winter to summer, only about one part in three hundred of the whole. However, in quantity of air this comes to, roughly, 10,000,000,000,000 tons which are shifted across the equator every six months from the warming to the cooling half of the world.

Not only is there a world-wide shift of the atmosphere with the seasons, but also there are like-caused, similar shifts of the air with the seasons between the oceans and the continents of each hemisphere. In winter the average temperature on the land is much lower than that of the adjacent oceans, and in summer much higher. Hence their air pressures also vary from season to season, being greatest over the continents during the colder season and least during the warmest, and over the oceans greatest in summer and least in winter. This difference between land and sea pressures varies widely from place to place, but on the average is greater than that between the summer and winter pressure of either hemisphere (northern or southern) as a whole.

The equatorial belt of low pressure varies a little in position owing to the annual shift of the axis of maximum temperature, or thermal equator—nearly always and everywhere a little north of the geographic equator. Evidently the shifting of this belt necessarily entails some change, though small, in the air pressure of the region covered by it. The high pressure belts also shift north and south, and through several degrees, thereby causing other annual swings in the pressures of many places. The Icelandic low, as previously mentioned, is much stronger in winter than in summer, while the Aleutian low exists only through the winter. Whether there is much change in the low-pressure belt around the border of Antarctica is not entirely certain, owing to lack of adequate observations through the year in that part of the world. Presumably, however, there are some annual changes in the depth and position of this belt also as there are in the intensities and locations of all the others.

In addition to these annual swings in position and intensity of the belts of high and low pressure, there also are marked occasional variations in the locations and strengths of their centers of action, the peaks on the ridges of high pressure, and the depths in the circumpolar valleys of low pressure. When marked shifts of these centers will occur can not yet be accurately foretold, nor indeed do we know in detail any specific chain of circumstances that produces such a result. We do know, though, that they often cause, so long as they last, a great difference in the weather of neighboring lands, and at times even of places far away. For this reason the centers of action, both the cyclonic and the anticyclonic, deserve and have received much study. We do not yet fully understand them, at least not so well that we can know in advance when and where they will occur nor how decided they will be, but that is no reason for giving over our study of them; rather is it a challenge to make our study all the more persistent and thorough. We should here emulate the Frenchman who, with the courage of his convictions, said that though we do not understand woman, and may never understand her, we are not for that reason going to give her up.

An even greater puzzle among the many pressure changes of the atmosphere are the occasional unheralded surges towards greater pressure or lesser pressure of roughly equal amount simultaneously, or nearly so, over an area of continental size. It may last for days and then disappear as mysteriously as it came. So far as the weather is concerned these surges appear to be of little importance, but until we do understand their cause and know their effects they will remain a taunt of ignorance that leaves us mentally ill at ease. Possibly, too, when we know them better we shall find them more important than we ever before dreamed they could be.

The most important of the pressure changes, due, we must remember, to changes in the amount of air overhead, are those that in middle latitudes occur on the average at least once a week, and often amount to fully three per cent of the whole. These changes are incident to the passage of those systems of wind, and their accompanying phenomena, which we call cyclones in which the pressure is relatively low, and anticyclones in which it is comparatively high. In most tropical regions, but not all, nor anywhere close to the equator, cyclones, often very intense, occur, but never an anticyclone. The anticyclone is the harbinger of fair weather, and the cyclone, wherever it occurs, the bringer of cloud and commonly also of rain or snow according to the current temperature. The forecaster locates on his map these systems of wind and weather, and from his understanding of the subject and intimate familiarity with such maps, quickly estimates what the weather will be a short time ahead, usually 12 to 36 hours, at each place over his territory, large or small. In some cases he makes a general forecast for at least a week ahead, and in others, as for the aviator, a very detailed forecast for only the next two or three hours.

There still are a few more irregularly-occurring changes in the pressure of the atmosphere that should be mentioned. (1) The precipitous drop in the pressure, amounting in some cases to probably one tenth of the whole, that occurs with the central passage of the destructive tornado. This extreme drop in the pressure is owing to the centrifugal force incident to the violent rotation of the air about the axis of the storm. (2) The irregular fluctuations of atmospheric pressure that accompany the earlier stages of a passing thunderstorm, due, in large measure, to irregularities in the downward velocity of the air in the front portions of such storms. (3) The minute fluctuations in pressure of 5 to 10 minutes period that sometimes occur when a shallow (a few hundred feet deep) surface layer of cold air is overrun by a wind of warmer air which throws the former, or cold layer, into waves or billows, much as winds make waves on the ocean. (4) Finally, if we think to have done with all these irregularities in the atmospheric pressure at one full swoop by the use of an automatic registering apparatus of exceeding delicacy, such as has been constructed and tried, we shall find that we have but multiplied our difficulties beyond endurance, for minute changes are continuously occurring that appear to be wholly without order or regularity of any kind.

But even all the above changes, annual and irregular, of the readings of the barometer do not exhaust its known variations. Indeed only a few years after the invention of the barometer in 1643 an article was published in the Philosophical Transactions of the Royal Society of London on the twice daily symmetrical swing of the atmospheric pressure, which in tropical regions amounts, from maximum to minimum value, to roughly one part in 150 of the total pressure. The amount of this pressure change decreases with increase of latitude, slowly at first and then so rapidly that it becomes uncertain beyond latitude 60°. Its maxima occur at about 10 o'clock, morning and evening, and its minima at 4 o'clock, both A.M. and P.M. There appears also to be a variation of the barometer with an 8-hour period. Furthermore there is a daily variation of the pressure from maximum, in valleys and other low places just before sunrise when the air is coldest and most contracted, to a minimum when the air is most heated and expanded. On mountain tops, and other places of considerable elevation, the pressure is minimum and maximum when down below it is maximum and minimum, respectively.

Still further variations in the surface pressure of the atmosphere are caused by the tidal actions of the sun and the moon. Those due to the action of the sun occur twice daily, but are very small, having a total variation even in tropical regions where it is most pronounced of only about one part in 40,000 of the whole pressure. It therefore can be detected only by averaging a long series of observations hour by hour so as to eliminate irregular variations and then subjecting the result to rigid analysis. As stated, there also are similar, but about two and a half times larger, variations in the pressure incident to the air tides caused by the moon. Even these no novice at the game would ever detect. Lastly—not the last that could be mentioned, but the last that here will be mentioned—there are the annual solar tides due to the swing of the sun north and south of the equator, wholly lost, perhaps, in the much greater thermally induced swings of the atmosphere in the opposite directions; and the lunar-monthly tides caused by the drifting of the moon from one hemisphere to the other across the equator. As just stated, not all the changes in atmospheric pressure, all the comings and goings of the air, have been mentioned in this chapter, but enough of them, no doubt, to make it clear that the ways of the air are devious and strange—full of puzzling problems that it is an intellectual delight to solve, and in every sense profitable to understand.

WIND

"Never mind the weather so the wind don't blow," runs the old song, turgid with the wisdom of experience, that our granddads used to sing in their boyhood days. But the winds do blow—blow comfort when gentle, annoyance when strong, and destruction when violent. Naturally, then, they were of vital importance to our ancestors of the tree tops, and important even to our great-great-great-to-the-nth-generation grandfathers who went to the other extreme and lived in caves, for whatever the shelter they found within the earth, surely the game that hunger compelled them to chase had to be captured on the outside in the realm of Boreas and all his blustery kin. Nor have the artificial caves we now make of all sorts of materials, from stone and mortar to stubble and straw, and place on the surface of the earth, improved the situation one whit, for they, too, even more than natural caves, leave us subject to stress and storm. There remains the same compulsion to leave them for the open betimes, nor are they always proof against every wind that blows.

And so it has come about that ages ago man asked: "What is a wind?" "Where does it come from?" "Where does it go to?" "What caused it?" They asked a lot of other questions too, but these will do for the present. To the first of these questions, "What is wind?" a Greek philosopher replied: "Air in motion." No better definition of wind than this ever was made nor ever can be—try it. The next questions proved to be more difficult. In fact until yesterday, relatively speaking, they were either given up wholly as unanswerable—"The wind bloweth where it listeth, and thou hearest the sound thereof, but canst not tell whence it cometh, and whither it goeth"—or else answered in an entirely fanciful and irrational manner, however pleasing the concept and beautiful the poetry that adorned it. According to one such concept a demigod, Aeolus, had charge of the winds, which he kept confined in a great cave, and from which this or that one was let out as whim, fancy or need might suggest.

A similar poetic concept was that of the personified north wind, Boreas, also a demigod, winged and wild with disheveled hair and beard, who likewise dwelt in a cave. Why he was supposed to dwell in a cave is not stated, but of course if, as believed, he was real and possessed of human qualities he had to live somewhere, and being superhuman that somewhere could not be like unto the home of a mere mortal, but rather some strange and mysterious place. Furthermore a cave, and there are many such, that has two openings at widely different levels gives out, on hot days, and from the lower entrance, a strong and continuous cold blast; and surely in the presence of such a weird phenomenon a personificationalist, such as a polytheistic poet, who knew nothing of its true and simple cause, might well attribute it to the blowing by mouth, as we ourselves would blow, of some giant, superhuman being.

Perhaps this is why Boreas occasionally was represented, not as winding a conch shell, as by the more aesthetic Greeks, mindful of the noise of the north wind, but with distended cheeks and rounded lips that urge the whistling blast. Similarly, and presumably for much the same reasons, it was common through the middle ages—the period of great faith and piety—to picture winds as coming from the puckered mouths of wingless cherubs.

The well-nigh universal ignorance of winds and their causes is further emphasized by the world-wide belief that by this means or that, and especially by some form or other of imitative magic, one could obtain any kind of wind he might desire, from calm to tempest, and of whatever direction. Nor were, neither are, these superstitions confined to recognized savages.

Ulysses is said to have received the winds in a leathern bag from Aeolus, and quite analogously, for centuries, in certain civilized countries, winds could be purchased, and were purchased, especially by sailors. Perhaps they still may be obtained at a few places, not in leathern bags, as Ulysses got his, for that would be inconvenient, but in knotted strings or handkerchiefs—one knot to be untied for a breeze, two for a blow, and three for a storm.

But our knowledge of winds has improved of late. We now can talk of the how and the why of them, and discuss their whence and their whither in the exact language of the physicist. We know that air has weight and inertia, and that like mass, or matter, of any other kind a force is required to put it in motion, or to change its speed; and also that when it is not somehow restrained a force acting on it gives it motion, or alters the motion it already has.

From this we correctly infer that every wind is caused by a push on the moving air. Furthermore, we know that this push is owing to the unequal distribution of the atmosphere itself over the earth, and that this vastly important irregularity is owing, in turn, mainly to the irregular distribution of temperature. This, however, is not the whole story, for there is mechanical reaction as well as action. That is, while pressure causes the air to flow the flowing air or wind causes a pressure. That a wind produces a pressure against any obstacle in its path every one will admit. He knows it from the bending, and sometimes breaking, of trees in the blast, and from the push he has felt on his own body when out in a gale. We also know the magnitude of this push in terms, say, of pounds per square foot of flat surface directly facing the wind, when we know the velocity of the wind and the density of the air.

But there also is another and very different force, or reaction, that comes into existence immediately the wind begins to blow, a reaction long unsuspected but of exceeding importance. It comes from the rotation of the earth, which causes an object moving horizontally over the surface continuously to deflect, or at least steadily to push, to the right (assuming one to be facing forward in the direction of travel) in the northern hemisphere, and to the left in the southern hemisphere.

To get some idea as to how this may be, imagine a cannon to be located at the north pole and to be aimed and fired exactly at New York City; let the speed of the projectile be one mile in three seconds, a fair cannon-ball velocity; and let it keep on its course a little above the surface of the earth, and wholly unaffected by it or the winds, until it gets as far south as the original target, or farther if this should be missed. And it will be missed, for during the time the projectile was on its way from the pole, the earth was continuously rotating from west to east at the rate of one revolution, counted by the fixed stars, in a little less than 24 hours, so that at the end of the projectile's flight of three hours, roughly, necessary for it to get so far south, not New York City, the original target, but some place in California, perhaps San Francisco, will be struck.

The projectile therefore in this case has gone or deflected far to the right of its initial course, as viewed from the surface of the earth. If, now, we would keep it directed all the time along the meridian through New York so that the object aimed at would also be hit, and, not missed a few thousand miles, we must arrange a tube, or groove of some kind, to carry the projectile and force it to keep up with the surface of the rotating earth. In this case the projectile and the western side of the tube or groove would press against each other—a force would be exerted on the projectile to make it keep up with the eastward movement of the earth, and of course an equal westward push would be exerted by the projectile against the groove. This is the deflective "force" of our projectile to the right (in the northern hemisphere) of its path. In reality the projectile merely keeps straight ahead on its course equatorward while the earth at the same time speedily turns under it. Thus the target runs away from the projectile and a wide miss instead of a hit is scored.

Obviously an exactly similar result, a miss, would occur no matter what the target shot at with our super-cannon from the north pole, or south pole either, only in the latter case the deflection would be to the left. Evidently, then, a wind across either pole and directed towards the equator would deflect-to the right if from the north pole, to the left if from the south pole—unless prevented from doing so by a suitable pressure against it.

All this is obvious enough for a projectile, or a wind, from either pole just mentioned, but what happens at the countless places away from these poles? Exactly the same sort of thing, only with less force, or smaller rate of deviation, because every point on the earth's surface, except those along the equator, is a pole about which the earth is turning with an angular velocity equal to that at the north or south pole times the sine of the latitude of the place. At the north or south pole it rotates once, measured by the fixed stars, in a little less than 24 hours; about any point 30° from the equator it turns once in just twice that time; and similarly for all other places.

Nonsense, we say, and it does seem so. But carefully suspend a good-sized metal ball by a wire from a point some distance above, as Foucault did under the dome of the Pantheon in Paris in 1851, and set it swinging in any direction and watch what happens. If the experiment has been carefully made the pendulum will slowly change the direction of its swing.

We can easily agree that if the pendulum were at the north pole and swinging towards New York City the earth would so turn under it that presently it would be swinging towards Chicago, and a little later towards San Francisco, and so on around until, when the earth had made one complete rotation, it again would be swinging towards New York City. We can also agree that the pendulum would behave in a similar manner at any place about which as a pole the earth was actually turning and only at such a place. Well, that concession has our logic hemmed in with no possible line of escape, for the Foucault pendulum behaves in just that way wherever it is mounted, save along the equator. Always and everywhere, except on the equator, the direction of its swing slowly changes; that is, in reality, the earth slowly turns around under it, at such rate that the time required to make the complete turn is equal to the time of rotation at the north or south pole divided by the sine of the latitude of the place of observation.

Every place then on the surface of the earth away from the equator is a pole of rotation, and everywhere the earth is rotating with a known angular velocity under the wind at that place. Hence every wind tends to its direction—tends to deflect to the right in the northern hemisphere and to the left in the southern hemisphere, and does thus deflect except in so far as by force it is prevented from doing so. It would seem too that at any particular place this force must be, and it is, the same for one pound of air as for any other pound of air going horizontally with the same speed. That is, this force is proportional to the weight of the object exerting it. It also, though that is not so obvious, is directly proportional to the horizontal velocity of the object moving.

This looks like complication enough, but there still are other forces that must be considered. If we fasten a marble or like object to the end of a string and then, holding the other end fast, whirl it around in a circle we shall find that there is a tension on that string—a pull from each end against the other that increases with the mass of the object whirled, the length of the string and the number of turns per second. Also, if in any way the string is shortened, by allowing it to wrap around one's finger, for instance, or otherwise, the rate of the turning will increase and the pull grow stronger.

Similarly, if water in a large vessel, such as a basin or a bath tub, has some rotary movement and then is allowed to flow out rapidly, the spin in the water will increase as it approaches the outlet until, in many cases, a hollow funnel of air extends from the surface even into the upper end of the drain pipe. Now precisely similar things, and for exactly the same reasons, happen also to the atmosphere. It flows hither and yonder over a rotating earth; rises here under local heating and sinks there owing to local cooling, and there is no rest for it, only ceaseless movement and turmoil. Also, making confusion more confounded, the mountains and valleys, hills, hollows, cliffs, trees and every other roughness on the surface of the earth, stir the winds into myriads of mingling eddies, surges and whirls whose only regularity is their utter irregularity, like a swarm of bees on the wing, only worse, for all the bees are of one size while the eddies are of all sizes.

But really the situation is not that bad provided we are willing to consider one question at a time, and willing also to leave some of the details for the amusement of our successors. As already stated, winds anywhere are the combined result of earth rotation and differences in pressure at the same level. Thus continuous heating of the air in the equatorial regions, such as always obtains, together with its ceaseless cooling in the higher latitudes causes it perpetually to overflow from the former to the latter and to underrun in the opposite direction. This looks simple and obvious, and would be if the earth were not rotating and if the heating and the cooling were continuous and uniformly distributed over their respective areas.

But neither the heating nor the cooling is uniform, and the earth is not still but rotating. However, since the tropical regions are all the time warmer than those near the poles we may assume, as a first rough approximation, that the heating in the former and the cooling in the latter are uniform. This, if the earth were not rotating, would lead to an approximately uniform and continuous overflow from the warmer to the cooler regions and a like return undercurrent from the colder to the warmer zone.

But, as stated, the earth is rotating and that leads to endless complications in the movements of the atmosphere, or winds. Everywhere friction between the surface of the earth and the atmosphere tends to bring the latter to rest with respect to the former—to stop all winds and everywhere produce a dead calm. Therefore air pushed to higher latitudes, where the distance around the earth is less, soon outruns the surface and appears as a west wind. Similarly, air on its way equatorward sooner or later is outrun by the earth and therefore become an east wind. Really this difference in distance around the earth along parallels of unequal latitude is only a part of the cause of east winds in equatorial regions and west winds in middle latitudes. The effect of the additional part, conservation of areas, is even greater than that which would result from the incomplete but obvious cause indicated.

From this it would seem that there should be a belt of winds from the east (east winds) along, and to a considerable distance on each side of, the equator, and beyond this belt another in either hemisphere of west winds. But such aroundthe-world winds would be their own undoing. They come into being owing, as we have seen, to the movement of warm and of cold air to higher and lower latitudes respectively. But such winds, if continuous around the earth would create along and near the interface between them an unbroken belt of high pressure parallel to the equator that would be an impenetrable barrier against any further interchange between the warmer air of the lower latitudes and the colder air of the higher latitudes, whereupon surface friction soon would bring all the air to a dead calm.

As a matter of fact these east winds and west winds, and the belts of high pressure between them, are more or less broken up and irregular. They are well developed and fairly continuous over the Atlantic, Pacific and Indian Oceans, but are only fragmentary and fitful over the continents. In great measure the tropical air makes its first movement towards the polar regions in the form of vast currents to the immediate westward of the ocean links of the high pressure belts; and, in more or less equal measure, the polar air moves towards the equator around the eastern ends of these same links of the high pressure belts.

Trade Winds

The tropical east winds that are over the oceans are called trade winds, in the sense of track winds, or more or less steady winds, along the same course. There are two trade winds on the Atlantic Ocean, namely, the North East Trade and the South East Trade. The first of these comes out of the north around the east end of that link of the northern hemisphere belt of high pressure that crosses the Atlantic Ocean. This wind as it flows southward soon is so outrun by the surface of the rotating earth as to become first a northeast wind and then more and more nearly an east wind.

Its velocity varies from around 6 miles per hour in October to about 12 miles per hour in April. Some of this air ascends, on being decidedly warmed, while some of it crosses the ocean and then turns first northwestward, then northward past now the western end of the same section of the high pressure belt by whose eastern end it had begun in flowing southward. Soon, as it moves northward, it outruns the surface of the earth and turns northeastward to eastward, thus merging with and becoming a part of the prevailing westerlies (winds from the west) of the middle latitudes.

The other trade wind of the Atlantic Ocean comes from the middle latitudes of the southern hemisphere, flowing first northward towards the equator past the eastern end of the Atlantic segment of the southern high-pressure belt, then, as the surface of the earth outruns it, turning northwestward and then nearly westward. Its velocity is roughly 12 to 15 miles per hour through the year. A portion of this South Atlantic trade wind also ascends, owing to surface heating, in tropical regions and flows off at high altitudes towards the adjacent pole, while another portion rounds the western end of the Atlantic segment of the southern high-pressure belt, after which it becomes a portion of the westerlies of the southern hemisphere. Exactly similar trade winds, only feebler as a rule, occur only the Pacific Ocean. Over the Indian Ocean trade winds occur only south of the equator. North of the equator monsoon winds, to be briefly discussed presently, predominate.

Countertrades

So much then for the "trades" which are the steadiest and most persistent of all winds. The next in order of persistence, year in and year out, and which in conjunction with the trades might be of decided advantage in transoceanic aerial travel are the counter trades, or return over-flow branches of the trades, resulting from their ascents in tropical regions. This ascending air flows off towards the higher latitudes of the same hemisphere in which their ascent occurred and after passing latitude 20°, or thereabouts, turn more or less eastward, so that the trade wind beneath and its overhead companion counter trade have very different directions. In this way the counter trades merge with and become portions of the normal or prevailing westerlies.

Westerlies

As already explained, owing to the rotation of the earth, any wind that has come from the general neighborhood of the equator and attained to latitudes beyond 20° begins to flow eastward, slightly at first and then, with increase of latitude, more and more nearly directly eastward. These winds are the westerlies (winds from westerly points) of middle latitudes. Though this is their average or prevailing direction they are by no means either as regular or persistent as the trades and the countertrades.

Polar Easterlies

Beyond latitude 60° in each hemisphere the surface wind again tends to blow from easterly points, just as it does near the equator. This is because of the flow of the cold air here away from near the pole to places where the distance around the earth along a parallel of latitude is greater, and therefore the eastward speed of the surface correspondingly greater—so much so indeed that the cold air is more or less run away from and thereby rendered an easterly wind. This polar east wind is far better developed over and around Antarctica, where the outflow of the cold air is more pronounced than it is over and around the Arctic Basin where the outflow is comparatively sluggish.

The winds above described, that is, the five trade winds, the five corresponding counter trades, the two westerlies (one in each hemisphere) and the two polar easterlies may be, and often are, called planetary winds, since they have their origin in two general or planetary conditions, namely, (1) rotation, with, in most places, a high surface speed, about an axis that is not far from normal to the direction to the sun, or, as astronomers say, to the plane of the ecliptic; and resulting therefrom, (2) a persistent great difference between the temperatures of the tropical and the polar airs that maintain an equally persistent interzonal circulation.

Monsoons

Another great class of winds, noted for their long duration and constancy of direction, are the monsoons. This word, of Arabic origin, means a season, and is here applied to winds of season-long duration. They occur in many parts of the world but are strongest, most persistent and best known in India and over the adjacent ocean north of the equator. Here, owing to the great contrast in temperature between the continent and the ocean, there is a persistent under or surface flow of air from the land to the water during the colder season, which flow, under the influence of the rotation of the earth, becomes a wind from the northeast—the northeast monsoon.

During the warmer season, on the contrary, the wind is prevailingly from the ocean onto the relatively warm land, and, as before, affected by the rotation of the earth which, in this case, causes the wind to come from the southwest. This is the southwest monsoon that brings the rains to India. The depth of the summer monsoon of India is about 3 miles, and that of the winter monsoon roughly 1.2 miles. Various other places over the earth have greater or less seasonal changes in the directions of their prevailing winds, owing to changes in the temperatures of the adjacent regions, but at no other place are these changes as pronounced, persistent and important as they are in India.

Land Breeze and Sea Breeze

Closely analogous to the monsoon, but vastly less in depth and extent and having only a diurnal period instead of an annual one, are the land breeze and the sea breeze, caused by the difference in temperature between the surface of the water and the surface of the land from the shore back some miles. They are most pronounced where the land is low and arid, skies prevailingly clear and general storm movements only occasional and commonly slight. The sea breeze, an on-shore daytime wind, is caused by the greater atmospheric pressure, at the time of its occurrence, over the water than over the land, incident to the lower temperature then of the surface of the evaporating water than that of the surface of the immovable soil. At night, especially when there are no clouds to send back radiation to the earth and keep up the temperature of its surface, the temperature contrast is reversed, the water surface then being the warmer of the two. Hence the direction of the horizontal pressure gradient, or push on the air, also is reversed and the surface wind turned back from land to sea. These winds are shallow, roughly 300 to 1500 feet deep, and seldom extend more than 25 miles inland. In most cases they are so graciously refresh- ing as to be as welcome as the family physician or parish priest.

Valley Breeze

Another daytime wind is the so-called valley breeze, a wind, commonly gentle, that on hot and otherwise calm days blows up mountain valleys. It seems to be owing essentially to the heating and expansion of the air over the nearby valley or plain, which heating increases the pressure in the air at a given height, more over the plain than over the adjacent side of the mountain (the air column to the level in question being shorter above the mountain than over the plain) and therefore causes a flow of the air towards and up the mountain sides and valleys.

KATABATIC OR DRAINAGE WINDS

A somewhat analogous class of winds, oppositely directed and ranging from imperceptible to severe, are the mountain breeze, canyon wind, glacier wind, and fall-wind, or, to use their recognized formal name, katabatic winds, of every sort. Another generic name for air movements of this kind is "drainage winds." In all these cases the wind is directed down grade along the surface. In the main it is a night phenomenon and owing to the cooling and consequent contraction and gain in density of the under air, incident partly to its own radiation, but mainly through contact with the relatively cold soil or other substance beneath—cold usually as a result of rapid loss of heat by radiation.

In many respects this air drainage is quite analogous to water drainage. It flows to lower levels, and from uplands runs into gullies, from gullies into drains, from drains into ravines and from ravines into canyons or valleys, and thence, it may be, out onto a plain of lower level. It may flow away gently, or ripplingly like a brook where the grade is moderate, and even plunge over a precipice in the form of an aerial cataract—a cataract, however, that in some cases never reaches the foot of the fall, because in the course of its descent the increasing pressure upon it makes it warmer and therefore lighter than the surrounding air which thereupon quickly arrests its rush to lower levels, and even causes a partial rebound. Drainage winds teach us many a useful lesson. The mountain camper, for instance, soon learns to pitch his tent above his evening fire, and not below it where the down-flowing air would pester him with the smoke. The Swiss valley dweller sets his house on a knoll so as to be above the nightly flood of cold air that surges around him. The wise orchardist selects for his trees a place part way up a slope, where the air by descent will have appreciably warmed, and yet above the lower levels where the cold air accumulates since it can drain away but slowly.

That is, he plants his orchard along the "thermal belt" where frosts, both early and late, are least likely to occur. The prudent aviator avoids landing at the mouth of a canyon when drainage winds can be strong, because they always are shallow and at times even violent. The cautious boatman when near a steep elevated coast of high latitude keeps his weather eye open for the williwaw, that abrupt, smashing avalanche of frigid air that seems to have sloshed over the mountain brim from the icy bowl beyond. The fisherman of Trieste, Fiume or Novorossiisk, to mention only the most conspicuous of such places, has a wholesome regard for the sudden and chilling Bora of hurricane force that, rushing from the snow-covered uplands pounds the sea into waves and the waves into a cloud of spray. The polar explorer, unless daring the winds to do their worst, as Sir Douglas Mawson seems to have done the year he spent at the "home of the blizzard," keeps away from steep slopes facing the sea.

Of course there are many other examples of gravity winds, as they also are called, or fall winds, another of their various names, but the above are sufficient to emphasize their interest, importance and world-wide distribution—always shallow and ranging from the gentle, balmy drift to a furious, frigid blast of hurricane force.

Tropical Cyclone

A radically different wind from any of those already mentioned, important, often violent and covering a somewhat circular or elliptical area, 50 to 1,000 miles across, is the tropical cyclone—the hurricane of the Atlantic Ocean, Gulf of Mexico, Caribbean Sea and eastern Pacific; baguio of the Philippines; typhoon of the China Sea; and the cyclone of the Indian Ocean. It is characterized by strong winds, up to 100 to 150 miles per hour, directed spirally inward and around a common centre, the place of lowest atmospheric pressure, clockwise in the southern hemisphere, counterclockwise in the northern hemisphere. It is further characterized by dense clouds, much thunder and lightning and torrential rain. It occurs over all tropical seas and oceans except the South Atlantic, but most frequently over the western North Atlantic, including the Gulf of Mexico and Caribbean Sea, Bay of Bengal and western Pacific, including the China Sea. It starts usually, if not always, in a doldrum or calm region, 5° to 20° away from the equator, and more particularly in that portion which presumably is a very sluggish atmospheric eddy—a section, 100 miles broad, perhaps, of air which though nearly quiet probably has a slight eddy motion owing to the presence on either side of it of oppositely directed "trade" winds.

Here the warm air becomes excessively humid, and for that reason, as well as because of its high temperature, comparatively light. Hence convection here becomes pronounced, clouds form and thunderstorms develop. This ascent of light air necessarily implies the corresponding inflow of surface air, and so it must happen from time to time that, owing to the rotation of the earth, and to the opposition in the directions of the winds on either side of the place of ascent, this incoming air is set into rotation, gently at first, but more and more vigorously as the centre is approached, like disturbed water as it runs into a drain-pipe at the bottom of a basin.

The ascent of the air causes it to cool by expansion, its cooling leads to condensation of some of the vapour present, condensation liberates much heat and this heat in turn continues and enhances the convection—not indefinitely, of course, but to a very considerable extent. Thus once begun this rotating disturbance is likely to perpetuate itself so long as a great quantity of water vapour may be had for condensation.

Hence, as the storm is captured by the major neighboring wind, commonly the trade wind on its poleward side, and often carried thousands of miles across a tropical ocean before reaching land, or being turned off to higher and cooler latitudes, its duration generally is at least several days, often a fortnight, and in extreme cases three weeks or more from its inception in some doldrum region to its last feeble efforts far away where water vapour is scarce and energy from its condensation adequate to the continuation of the storm therefore not available.

Commonly it travels a long way from east to west in a trade wind stream, then slowly turns around the western end of the adjacent link of a high pressure belt and swings off to the east again, but at much higher latitudes and with properties largely different from, and more complicated than, those it formerly possessed. In many cases it is driven onto land where, for a time, it brings heavy rains, and then, as a rule, fades away.

The rate at which these storms use energy—their horsepower—is amazing. Thus from the exceptionally large amount of excellent data available about a particular typhoon the Japanese savant, Horiguti, computed that the latent heat of condensation was freed at a rate equivalent to 1,200,000,000,000 horsepower, and that the rate of its mechanical dissipation of energy was equal to 27,000,000,000 horsepower. Yet some people talk seriously about destroying the typhoon, or hurricane, by dropping bombs onto it from aeroplanes. How the aviator is to reach his vantage point over such a storm is, of course, his problem. But there is no occasion to worry about the details, for saying "boo," to a blizzard would be in about the same proportion and quite its effective in stopping it.

Extratropical Cyclone

The great contrast in temperature maintained between the polar and equatorial portions of the earth keeps the atmosphere, as previously explained, on a continuous round of circulation between them, because the equality of temperature this circulation tends to produce never is even approximately attained. And because this interzonal circulation is over a rotating earth it follows, as may be demonstrated mathematically, that in mid-latitudes it must occur, and it does so occur, largely in the form of great swirls between broad currents of warm air moving poleward and cold air flowing equatorward. Furthermore, since the warm air moving to higher latitudes outruns the earth it therefore must be to the eastward of its companion cold current flowing equatorward and thereby tending to lag behind the rotating earth.

Hence the swirl thus produced between these two currents, the place where the air tends to run ahead on the eastern side and to lag behind on the western, is a region of atmospheric deficiency, or of low atmospheric pressure, with winds directed round about it counterclockwise in the northern hemisphere, clockwise in the southern, and, in each case, turned spirally inward. The simultaneous directions of the wind all around the centre of this low pressure, or place where the pressure is least, seem, on first considerations, to indicate that the air is spinning spirally around and towards that centre—around and around St. Louis, for instance. But St. Louis is fixed in position, while the centre of the storm is not, but moves on eastward 20 to 30 miles an hour, as a rule, and so one gives up his first inference that St. Louis is the hub of the winds and jumps to the next "obvious" conclusion, namely, that they spiral around and around the traveling centre like the tire of a rolling wheel about the axle. Well, that is wrong, too, for the tire and the axle do not keep pace with each other. The storm is not a spinning mass of always the same air sliding over the earth, but a traveling swirl of constant type but of ever renewed material. Well then, if both these inferences are wrong, what course does the air follow in such a disturbance? We have learned something about that from the drift of clouds and of free balloons, and from the state and condition of the air at successive times and places, and know now that this course is not by any means so simple as we once thought it was.

But we must not let this fascinating flight of an individual air particle develop into a red herring dragged across the course of our subject—wind.

In the area of an extratropical cyclone, roughly elliptical and ranging from 200 to 2000 miles across, there always and everywhere is wind, usually of 10 to 30 miles an hour, but by no means confined within these limits. The cyclone is of frequent occurrence, two or three to a dozen a month, over most of the earth beyond latitudes 30° N. and S., and most of the winds of these regions are associated with them. As previously explained, the extratropical cyclone is a product of the interzonal exchange of air over the surface of a rotating earth. It therefore consists of two distinct parts, a warm and humid eastern portion where the air is of tropical origin, and a relatively cool and dry western section where the air has come from a higher latitudes. Wherever the boundary, 100 to 1000 miles long, between these two great counter currents of air passes, as it moves over the earth with the storm, there and then the direction of the wind quickly changes, from southwest, likely, to northwest, and its quality from relatively smooth-flowing to gusty. This traveling interface between the two wind systems, the tropical and the polar, is known as the wind-shift line or the polar front. It brings showers of the so-called clearing-up type, many of which in the summer are accompanied by thunder and lightning and squally winds.

A great deal more could be said about the winds of the extratropical cyclone, a disturbance that can be started in three or four different ways, but that would involve tedious details which, except for the specialists, would tend to obscure the major facts here stated and general outlines just drawn.

Anticyclone

The anticyclone, a system of winds over a region where there is more air per unit area, and therefore higher atmospheric pressure, than there is round about, is entirely an extratropical phenomenon, roughly elliptical in shape and commonly 200 to 2000 miles across, which, like the cyclone of the same latitudes, may be originated in more ways than one. Owing to the relatively high pressure within its area the lower air flows outward from an anticyclone, but of course under the influence of the rotation of the earth.

Its direction therefore is spirally outward from the centre of highest pressure, and around it, clockwise in the northern hemisphere, counterclockwise in the southern. But as in the case of the extratropical cyclone the direction of the wind in an anticyclone does not mark its course. Near the centre of the anticyclone the winds always are light, while close to its border they are much stronger, sometimes approaching or even reaching gale force. There is, however, a maximum possible velocity which the wind of an anticyclone can have, while no such limit applies to the winds of the cyclone. Another interesting contrast between these major disturbances is the fact that in the cyclone, especially the true cyclone of the tropics and subtropics, the swiftest winds are near the centre of the storm-not at the centre for there the winds often are still, but within 5 to 15 miles of it, while the swiftest winds of the anticyclone are near its border. On the average the velocity of the wind in an anticyclone is only 5 to 15 miles per hour, and distinctly less than that of the cyclone of the same region and season.

One of the most important and striking weather phenomena of the anticyclone is confined to the winter months and is most frequent and severe in the central United States. This is the cold wave that rushing from high latitudes sometimes carries freezing temperatures even to the border of the subtropics. It brings also the dread "norther" of Oklahoma and Texas, and the true blizzard of the Great Plains whose howling, frigid blasts, filled with powdered, blinding snow, give no quarter to man nor beast that ventures forth or is found afield.

Tornado

Child of the cyclone and the anticyclone, of the south wind and the north wind, where brought into closest propinquity, and born among the clouds, is that meteorological *enfant terrible*, the tornado, the smallest yet most vicious and destructive of all storms. It may be, and generally is, only 1000 feet, or less, in diameter, yet the velocity of its mad, dancing—dervish winds as they wildly spin around and up the central vortex may be 100, 200, even 500 miles an hour. So exceedingly violent is its action that few things can withstand the fury of its blasts. Like one of its parents, the cyclone, its gyration always is counterclockwise in the northern hemisphere and clockwise in the southern, and like its other parent, the anticyclone, it never comes near to the equator. The area covered by this storm is rather small since the length of its path commonly is less than 25 miles. Also its occurrence is, at most, only occasional, hence despite the terror and destruction it works wherever it does hit, the chance of being hit by it at all is so slight that relatively few people, even where it is most likely to occur, show any dread or concern about it.

Waterspout

A usually much smaller and less violent whirling storm than the tornado is the well known waterspout. This is not a rotating column of sea-water, 1000 feet high, as many seem to think, but just a column of cloud, formed in the humid air by its cooling due to the expansion under decrease of pressure incident to rapid rotation. It may be a tornado at sea, in which case it has the same restrictions as to sense, or direction, of rotation, and the same inability to approach the equator. On the other hand it may start from the surface, owing to the local ascent of air that is very humid and therefore relatively light, just as dust whirls are begun by the concentrated ascent of highly heated desert air. Waterspouts of this origin, like their desert brothers, the dust devils, as they often are called, occur at practically all latitudes, even on the equator, and wherever they do occur rotate in either sense, clockwise or counterclockwise. Also in whichever sense a waterspout or a dust whirl starts rotating, that way it continues to rotate as long as it lasts.

The duration of the waterspout commonly is very brief, 15 minutes, or thereabouts. It therefore is an excellent thing with which to prove the efficacy of magic; a little whistling, for instance, or a threat by the pointing of a sword, and it soon vanishes—as it was going to do anyway. Or, if one is more naval-minded, he can fire a gun at it—cannon to popgun, it makes no difference. If a gun is not convenient point a stick at it, and if we haven't got a stick our finger will do quite as well, for whatever we do or don't do the spout soon will be only a thing to remember and talk about.

THUNDERSTORM WINDS

Another special class of winds are those that normally accompany the thunderstorm. Commonly this storm is preceded a few minutes by a gentle wind directed towards its nearest portion. This is abruptly followed, just before the rain begins, by a much stronger and distinctly turbulent wind in exactly, or nearly, the opposite direction. The first of these is the indraft of surface air, on its way to great heights where the vapour it carries, or is a part of it, is to condense to cloud and rain and the rain in turn to play its important role in the production of thunder and lightning. The second, or outflowing turbulent wind, has come down from above, owing to increase of density due to cooling, and is swift because it has retained the faster velocity winds usually have at considerable heights.

Eddies

Sometimes one is surprised when, on climbing a mountain with a strong wind at his back, he finds, on going down the opposite side, a wind in his face. He is puzzled to know how the wind can blow up the mountain on both sides at the same time. The mountain does the trick.

As the wind crosses the crest it tends to go on at the same level, but it also catches up some of the air beneath it and drags that along, just as the air in the tube of an atomizer is pulled out, and with it some of the liquid beneath, by the blast across its top. This depletion of the air between the side of the mountain and the sheet of crossing wind, or start towards the production of a vacuum, necessarily leads to an inflow of an equal amount of air from outside, and the most available source is down the mountain. Hence a wind crossing a mountain range induces a return eddy current up the lee slope that is a great puzzle to the uninitiated, and great worry to the student of physics who tries to explain it with all the mathematical and other details satisfactory to an exacting professor.

Foehn

Sometimes the wind that crosses a mountain is so pulled down on the lee side that it strikes the valley or plain beyond, now gently and again even violently. As this crossing air descends to lower levels its temperature increases owing, as we know, and in proportion to, the gain in pressure on it, or amount of air above it. If, now, there is no cloud or rain on the mountain this gain in temperature by the descending air on the lee side is only equal to its cooling in the course of an equal ascent on the windward side, and nothing strikingly out of the ordinary happens.

When, however, abundant rain does fall from the ascending air the cooling is much less, owing to the latent heat of condensation freed, or rendered sensible, than it is when the air is dry, while the amount of warming is substantially unaltered, the air on the lee side of the mountain being clear, or nearly so, in each case. On such occasions the descending air reaches the valley or plain abnormally warm and dry;

indeed so great are these effects that now and again they even change fields of deep snow into a dusty plain in a few hours—yes, hours, not days.

TIMES OF GREATEST AND LEAST WINDS.

Nearly everywhere over land, but no much so over the ocean, there are diurnal changes of the wind velocity. Over level regions the velocity of the wind is least a little before sun-up, when the surface air is coldest and convection least; that is, when surface friction has acted longest on the same identical air and brought it nearest to rest. Similarly the winds are greatest, on the average, when convection has effectively brought upper, and swiftermoving, air most abundantly down to the surface, commonly 1 to 2 o'clock P.M. On mountain tops the hours of maximum and minimum wind velocity are just the reverse of those of the plain.

Also the winds are stronger in the winter, on the average than in the summer. This is because the push, or horizontal pressure gradient, that makes the winds, is greater, as a rule, in winter than in summer, and this in turn is because the change in temperature with latitude, which largely determines the corresponding change in pressure, is greater in winter than in summer.

Wind Velocity and Height

Another interesting and exceedingly important thing about wind is the fact that generally its velocity changes a great deal with change of height. Many studies have been made of this phenomenon, though some of the best of them are pretty well removed from the view of the average individual by, to him, an opaque wall of mathematics. This seclusion was not intentional, but unavoidable owing to the nature of the subject. It is easy to observe what happens, but the details of why it happens are, in many cases, quite difficult to explain.

Whenever there is an appreciable wind the roughnesses on the surface of the earth throw the lower air into a turmoil of rising, falling and mingling eddies. Hence, as the surface is the place where the friction occurs, the velocity of the wind must at first rapidly increase with increase of height, and also progressively change in direction to the extent of 30 or 40 degrees, owing to the rotation of the earth. In many cases this increase of velocity continues, but at a slower rate, up to the level of the highest clouds.

After this the velocity falls off until at twice that level, roughly, it again is no more than at the surface. At still greater heights the velocity once more increases. Often, in the forenoon especially, the increase of wind velocity with height continues up to the 1,500 foot level, or thereabouts, and then decreases until at twice that height the velocity has fallen to surface values. Beyond this level it again increases at a fairly uniform rate, and all the rest of the way behaves as above explained. Occasionally, on the other hand, the air is practically calm from the surface up to great heights—5 or 6 miles at least, and perhaps higher.

Chapter 5

Forms of Precipitation

The meteorologist borrowed the term precipitation from the chemist and then generally gave it a meaning different from that used by its original proprietor. The chemist means by it the process of separating a substance from its solution in such form that it settles or tends to settle under the force of gravity. The substance itself he calls a precipitate.

The meteorologist sometimes makes this distinction too, but more commonly he means by precipitation not a process but a product, such as rain or snow. The chemist gets nearly all his precipitates from solutions, and commonly, too, as a result of some combination into a new substance. The precipitation (precipitate) of the meteorologist, on the contrary, is not a new substance it all, but the same old substance in a different state, that is, liquid water, such as rain or solid water, as snow, for instance, condensed from gaseous water. Some of the chemist's precipitates settle quickly, while others fall slowly, and a few take forever-and-a-day to come down. So it is also with the meteorologist's precipitations. Large hailstones fall with a speed that is dangerous, drizzle drops just dawdle along, and dew doesn't fall at all—it is just said to fall.

There is a tendency on the part of many meteorologists to confine the term precipitation to only that water, liquid or solid, which, after dropping out of the clouds, actually reaches the earth. In its broader usage, however, it includes, as explained, all natural condensates from gaseous water to liquid or solid water. It is here used in that more comprehensive sense.

FOG

One of the most interesting, and in some respects an exceedingly important, form of precipitation, is fog, a great swarm-like assemblage in the surface air of hundreds of thousands of droplets per cubic inch so minute that it would take 7,000,000,000 of them to make a teaspoonful of water. It is a surface cloud whose myriad myriad droplets have formed on condensation nuclei, such as ultra-microscopic particles of seasalt, incident to the cooling of the air below its dewpoint. In many cases this necessary cooling occurs on still clear nights in river and creek valleys, as the result, chiefly, of the loss of heat by the surface through radiation.

When the sky is overcast with low clouds their radiation to the earth is nearly equal to that of the earth to them, and the cooling of the surface therefore, and, through it, the adjacent air, insufficient to induce condensation.

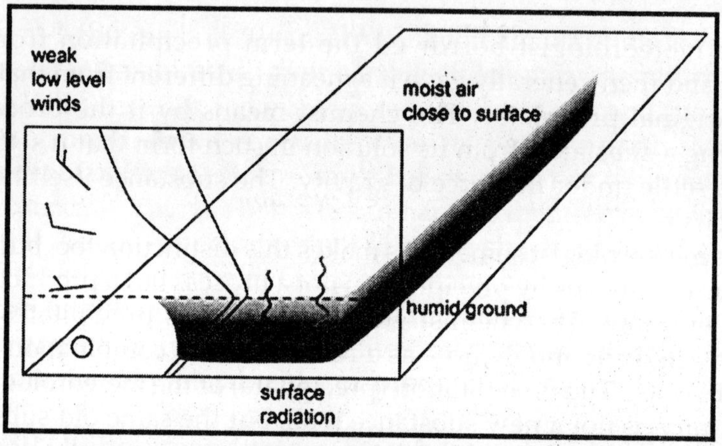

Fig. Radiation Fog

Similarly this type of fog, radiation fog as it commonly is called, can not and does not form, even on clear nights, when there is a considerable wind, as the cooling then also is slight. The loss of beat by radiation is not, of course, affected by wind, except indirectly through its effect on the surface temperature; but the fall in temperature and the depth of the chilled air, and hence the formation of fog, are affected by it.

Thus, whenever a dead calm prevails the cooling surface of the earth cools in turn only a very shallow layer of air, practically just that which is in direct contact with the chilled objects on the surface, since it itself is a very poor conductor of heat.

This chilling of the mere surface air, no matter how pronounced, does not produce fog, a sort of cloud above the surface, but only dew or frost condensed from the air that is chilled by the cold objects with which it is in direct contact. If there is a little wind, just enough to mix the air through a layer 10 to 50 feet deep, the cooling, though less than when there is no wind at all, often is enough to push the temperature of this mixed air below its dewpoint and thus to fill it with fog.

That is, the considerably cooled surface air, by mixture with a rather small amount of air above, cools it in turn below its point of saturation and thus induces throughout its volume condensation on the ever present, innumerable, hygroscopic nuclei, in droplets too small to fall to the ground against the diffusing force of the existing slight turbulence caused by the slow wind movement.

When the winds are strong, however, the consequent pronounced turbulence brings, one after another, all portions of the air through a considerable depth into contact with the surface and thus the loss of heat, though actually greater than it is on still nights, owing to higher surface temperatures, now is shared by a far larger mass of air and the fall in temperature, or actual cooling, thereby much reduced, and the occurrence of fog prevented no matter how bright the stars. In short, even when the skies are clear, no wind at all implies no fog; very light wind, say 2 to 6 or possibly 8 miles per hour, fog, if other conditions are favorable; strong wind, again no fog.

Fog is formed also by the drifting of warm humid air onto a cold surface off Newfoundland where the air from the Gulf has come over the much colder Labrador Current. It forms also, though to a much less extent, where the air is distinctly colder than the water. This is an exceedingly common occurrence on calm cold mornings over ponds, lakes and streams, from which one often sees wisps of fog rising at numberless places like so much steam from hot water. That is why the usual name for this phe- nomenon is steaming. It is only the condensation of the moisture in the humid surface air as it is buoyed up and mingled with the colder air above.

Ordinarily the evaporation is so slow, and the air above so dry, that this steam-fog is only a shallow, gauzy affair of but little consequence. Occasionally, a in high latitudes, the air is so very cold that the "steam" immediately changes to ice needles, or frost smoke, as it usually is called.

Not only may fogs be produced by the drifting of warm humid air over cold surfaces, and by the importation of cold air over warm water, but also, however and wherever formed, they themselves may be transported by gently moving air to other places, even in many cases to localities at which fog seldom is generated in situ. Such advection, or drift, fogs are exceptionally frequent just inland from mid-latitude seacoasts along which the prevailing wind is on-shore.

In certain cases this moving fog is largely caught by trees, notably by the towering redwoods of California, as it filters through them, and thereby condensed into a fog-drip that in every particular is the equivalent of just so much gentle rain. And this rain is so frequent too in various places as to maintain luxurious vegetation where otherwise only the spiny, stunted growths of an arid region could survive.

CLOUD

In general, a cloud, a thing that every one knows when he sees it, though few know much about it, is essentially just a fog in the sky, though on the average its droplets are roughly 35 times as bulky as those of true fog—so bulky indeed that if they were merged together a teaspoon could hold only a paltry 200,000,000 of them without running over. But not all clouds consist either wholly or even in part of these droplets. Some contain also rain drops, each of which is equal in mass to many thousands of the average cloud particle, while many others contain no droplets at all, but only snow in some one or more of its many forms, from discrete, tiny columns or needles to matted tufts an inch or more across.

Clouds differ from each other not only in respect to composition, some consisting of water droplets and others of ice particles, as just stated, but also in height above the surface, appearance and mode of formation. And these conditions are the criteria according to which we classify clouds and name them—a delightful intellectual diversion for which our forefathers felt no need and never practiced. When they had occasion to speak of a cloud they generally just called it a cloud, and let it go at that, without specifying what sort of a cloud it was. In fact it was not until 1801 that clouds were christened at all in any European language, and until 1803 that they were given names that proved to be generally acceptable.

A second edition of this international cloud atlas, only slightly modified and extended, was published in 1910. In general this atlas was quite satisfactory and became widely known and extensively quoted in books and periodicals. In 1922 another International Cloud Commission got busy and at the end of ten years produced a handsome work that has much to recommend it, especially its many excellent cloud pictures. But shade of Luke Howard! How some of the good old definitions have been changed, and how unwieldy for practical use, are both it and the abridged form that appeared in 1930! Many of us expect another International Cloud Atlas to appear some day in which the revisions largely will be reversions. Also there is urgently needed a working cloud atlas, cheap and above all convenient to use.

All the more commonly recognized clouds are listed below, though the definitions and comments are less formal than those given in international atlases.

Cirrus

The highest, generally, of all clouds, with the exception of two rare forms, to be mentioned later, the natures of which are unknown, is called "cirrus," and was so named by Luke Howard, because of its frequent appearance in the form of swirls and ringlets. It is a detached, white cloud, fibrous in structure and too thin to cast a distinct shadow or blur the outline of the sun or moon. Owing to the great heights at which it usually occurs, 6 to 7 miles above sea level in mid-latitudes and half as high again near the equator, its temperature commonly is very low, -50° to -80° Fahrenheit, roughly. And because of this very low temperature the citrus cloud nearly always, though perhaps not quite invariably, consists of ice particles.

Although the cirrus usually is higher than any other cloud, it does not always have that topmost position. Indeed it can and does occur at any level within the convectional portion of the atmosphere-the troposphere—from its highest reaches clear down to the surface of the earth. It occurs now and then beneath a cloud sheet of a different type and even between two such sheets or layers. Furthermore, it assumes as many different forms as Proteus, to many of which distinctive names, seldom used, have been given. When it slowly disappears it is a harbinger of fair weather, but when it grows denser it frequently is the forerunner of a general rain—a sort of a pilot or scout for the ranks of heavier and lower clouds following in its wake.

Cirro-Stratus

The next highest cloud also is white and commonly too thin to blur the outline of the sun. It may appear as only a formless milky haze or as a more or less continuous, fibrous sheet or layer. It therefore has been given the appropraite name "Cirro-Stratus," that is, a stratum of cirrus cloud. It shows the well known large rings or halos around the sun and the moon, from which we know definitely that it consists of minute ice crystals, as they alone so refract the light of the sun or moon as to produce these circles. When this cloud tends to thicken there generally are still denser clouds on the way bringing rain or snow in 6 to 24 hours.

Cirro-Cumulus

Sometimes the cirrus, or cirrostratus, seems to curdle, as the housewife says of milk when it coagulates into lumps. Such a cloud canopy, consisting of small lumpy clouds, often with more or less fibrous edges, but practically without shadow, and irregularly distributed, frequently is called a "curdled sky." Usually, however, the cloud lumps are arranged in orderly rows, or waves, that remind one of the colour patterns on a mackerel's back. This is the "mackerel sky."

Alto-Stratus

The next cloud type in the order of decreasing altitude is the alto-stratus, a heavy, whitish to bluish veil of clouds, sometimes fibrous in texture and again structureless, which we call alto-stratus. Commonly it is thin enough to reveal the approximate position of the sun, as if seen through ground glass, but too thick to show either coronas or halos.

It may be formed in any one of several ways, chiefly by the spreading out by wind of the upper portions of cumulus clouds; by the cooling in place of a layer of moist air; and by the flow of a layer of warm humid air over colder air, the cloud forming along the interface. It is not disturbed by numerous local convections as we know from the fact that they would make of it an alto-cumulus, a cloud it does not resemble; neither is it a product of a wide-spread active convection, since that would cause a more or less steady rain or snowfall, which is not a feature of the alto-stratus.

It is recognized that the above definition of alto-stratus is in keeping with that given by the older international cloud committee, appointed in 1891, and not with that of the committee of 1922. According to this later definition the altostratus always is fibrous, at least in parts. If so, then the fairly common "ground-glass" cloud of the same level and density that shows no such structure is without recognition. In this case many of us prefer our old friend to the new acquaintance.

Alto-Cumulus

Somewhat lower, on the average, than the alto-stratus is the alto-cumulus, a canopy of closely packed but isolated or nearly isolated cloud masses, seldom more than a few diameters of the sun or moon across and nearly always arranged in more or less conspicuous rows, often in two directions.

The smaller of these cloud masses are white; the larger appreciably shaded. When these clouds are slowly evaporating, and are 20° to 30° from the sun, they often afford wonderful colour displays—especially the delicate pinks and greens that characterize the finest mother-of-pearl. These exquisite patches of colour are but fragments of coronas produced by the diffraction of sunlight by water droplets. This iridescence in addition to charming us with its beauty, tells us also that the alto-cumulus consists of water droplets and not of ice crystals.

Droplets produce coronas, iridescence, and rainbows; the crystals give the many forms of the halo.

The alto-cumulus, like the cirro-cumulus, appears to be caused by local convections in an unstable layer of air, but one of greater moisture content than that in which the cirrocumulus forms. When the quantity of the humidity present is unusually great an isolated cloudlet here and there grows larger and much taller than its neighbors. In this case the cloud sheet may be called alto-cumulus castellatus, and is so called when are especially discriminating. Such a cloud not infrequently is a forerunner of rain.

In practice, it is the whole cloud sheet that is called altocumulus, and not one of its individual parts to which, ordinarily, no name at all is given, though actually it is a cumulus.

Strato-Cumulus

There is yet another cloud sheet of irregular density, a sheet of largish, dark grey balls or rolls of cloud that frequently are united by thinner cloud. These masses are larger, denser and commonly at a lower level than those of the alto-cumulus. Also in in many cases, especially just after the passage of a general storm, this cloud has the form of a continuous canopy in irregular folds and layers. Another type of this cloud is low and fluffy in appearance, and often, presumably, produced by the convection due to turbulence.

The tops of the unit masses of this cloud often are raised and rounded, showing that they are produced each by a local convection, while their bases have an approximately common level. In many cases, however, the tops of the separate masses spread out, like smoke under a ceiling, until they nearly or quite merge together. And they spread out because they too, like smoke in a room, have reached a ceiling—the limit to which the distribution of the temperature in the atmosphere permits them to rise.

Nimbus

When a general, or cyclonic, storm is passing the clouds associated with it often run the changes from cirrus, as the most advanced scouts, to cirro-stratus, and then an altostratus that grows denser, darker, and lower, covering the entire heavens and without distinctive form, until finally rain or snow falls from it steadily. In this final state, and even before that, so long as rain or snow seemed imminent, this cloud sheet formerly was called nimbus, and that is what some of us still want to call it who see no need whatever for giving it the new name "nimbostratus" adopted by the latest cloud committee. The two names refer to one and the same cloud, and therefore the new and longer name can only take more time to write, more space to print and cause more cloud confusion, a confusion that already was so great as to shake ones confidence in many, if not most, cloud observations.

The nimbus cloud normally results from the gradual ascent of humid air over an obstacle in its path, rising land in some cases but more commonly cooler air in front of the storm centre. Some rain or snow can and does fall from nearly, if not quite, every type of cloud, and even, on rare occasions, out of a clear and cloudless sky but that does not justify calling anything nimbus except the cloud just described, from which rain or snow normally falls in appreciable amounts.

Cumulus

When the weather is calm and warm and the sky generally clear, surface heating by sunshine starts rising many a column of air here, there and yonder, and when that air is moderately humid a cloud forms at the top and in the upper portion of each well-developed pillar—a cloud rounded above, with bulging sides, like a great pack of wool or a giant head of cauliflower, and flat at the base. This is the cumulus cloud, brilliantly white where the sun shines on it, and blue-grey to dark over the shaded portion, except along the edge not far from the sun where there is a silver lining—a wondrous capital of marvelous carving supported by an invisible column. Some of these clouds tower higher than their neighbors, but the bases of all, big and little, at any particular time and place, have substantially the same height, the height at which the current surface air on rising will have cooled to its dewpoint, or temperature at which saturation is attained and condensation begins.

During its growth the air within a cumulus cloud is in a state of vigorous agitation, as is evident from the way its sides and top continuously change in size and shape, and confirmed by the jolts and tossings the aviator experiences when in a large cloud of this type. From the origin of the cumulus cloud, namely, vigorous local convection, incident to surface heating, of moderately humid air, one correctly infers that in middle latitudes it is essentially a summer phenomenon; that it is most numerous in tropical regions and rarest near the poles; that at times it forms a chain of balloons, as it were, in the sky that delimits the course of a river, or a seashore, below; that over land it is most numerous by day and over the ocean at night; that it frequents islands far more than the open ocean; and that, among many other things, it is a fair weather phenomenon, since it can not occur when the sky is generally overcast, as adequate surface heating does not then occur, nor at the time of any considerable wind, since that would prevent the establishment of large columns of ascending air.

Cumulus-Nimbus

When a cumulus cloud grows to great size and height by the continuous feeding into it from below of humid air, or has been formed along a squall line by the juxtaposition of warm and cold air, the time comes when for one reason or another vast numbers of the droplets become so large and heavy that they fall out as so much rain which commonly reaches the ground, though in arid regions it sometimes evaporates wholly in mid air. As soon as this rain phase of the cumulus cloud begins we heretofore have been ready to change its name to cumulo-nimbus.

As the rain begins thunder and lightning also usually occur. In many cases also the cumulo-nimbus spreads out at the top into a more or less well-defined anvil shape. Finally, the topmost portion of the cloud frequently, but by no means always, nor at first, is combed out by the wind into a sheet of cirrus. And yet this is the one phenomenon by which the 1922 Cloud Commission distinguish between the cumulus and the cumulonimbus. They say: "Masses of cumulus however heavy they may be, and however great their vertical development, should never be classed as cumulo-nimbus unless the whole or a part of their tops is transformed into a cirrus mass."

Again: "None of these cumulus clouds should show ice crystal clouds (hybrid cirrus) at their tops; this would mean that they had reached the cumulo-nimbus stage." And in the formal definition of cumulo-nimbus there is this phrase: "... the upper parts having a fibrous texture." Finally, their brief definition of cumulo-nimbus is: "Cumulus clouds of great vertical development, with the tops composed of ice crystal clouds."

Well, if we must accept this definition of cumulo-nimbus, then this is the definition of cumulo-nimbus we must accept, being both official and international. But, for all that, it leaves some of us a little confused in our minds, as we say-those of us who have seen many a heaped-up cloud from which abundant rain was falling and in which there was frequent and vivid lightning, but of which no portion was cirrus nor showed any signs of cirrus. We called such a cloud cumulonimbus, and we very much dislike now to call it just cumulus. Indeed, rain is the really important differentiating feature between cumulus clouds, and not the trifling circumstance of the presence or absence in their uppermost portions of a little cirrus. And our confusion is greater, not less, when we recall other heaped-up clouds, cumuli we then called them, and still would like to call them, which developed abundant cirrus, but never furnished a drop of rain or a flash of lightning.

This is another case where some of us much prefer our old and faithful friend to the new and untried acquaintance.

STRATUS

Sometimes the sky is covered with a comparatively low layer of grey formless cloud from which little or no rain falls, only a drizzle at most. This cloud is called "stratus," and the height of its base varies from just free of the surface to about 3000 feet. It may be formed in several ways, such as the flow of warm humid air over colder air beneath, the cloud forming at their interface; by the drift of fog over relatively warm land, where the lower portion of the fog is burned off, leaving the upper portion as a stratus cloud; and by the relatively shallow turbulence convection induced by surface friction and surface obstacles.

Stratus is the lowest and most nearly uniform of all clouds. Indeed it is so devoid of contrasts that it scarcely can be shown in a photograph at all except where it is pierced by a hill or mountain, and even in that case a lively imagination is a great help—not to supply the details, for of these there are none, but to awaken that lonesome, dreary feeling one has on a day of drizzle. It would seem that, in all conscience, Nature had completely exhausted her sartorial ingenuity in the making from vapour alone of all the above sorts and kinds of cloud robes. But here too her ability has no bounds. There were scraps left over, as always, when the robes were fashioned.

BILLOW CLOUD

Although the billow cloud is only a special form of the alto-cumulus, it is so strikingly different from the average cloud as to arrest attention and to call for a distinctive name. It occurs in long, narrow and evenly spaced strips of white cloud. Even the individual strips themselves often show considerable detail of structure. It generally is supposed that this cloud occurs on the crests of waves or billows formed by the flowing of one current of air over another.

However, while the name "billow cloud" generally is restricted to a particular form of the alto-cumulus, the same sort of structure though of much finer grain, occurs as a frequent type of the cirro-cumulus, and likewise a far coarser one as occasional variety of the strato-cumulus. As just implied, two sheets of air of different density flowing horizontally the one over the other develop waves or billows at their interface which billows of course are coolest in their highest portions or crests and warmest in the troughs.

Hence when the humidity is just right such billows are cloud-crested and clear-troughed. But according to the theory of these air waves the distance from the crest of any one to that of the next is far greater than that from cloud to cloud in the delicate cirrocumuli ripples, and greater too than the distance between the rows of a wave-like alto-cumulus or billow cloud. Hence, while actual air waves, induced in the manner explained, probably are the cause of certain groups of cloud strips of coarse grain, or long wave length, it seems probable that the beautiful alto-cumulus billows are not produced in this manner, and certain that the citrus ripples have some other origin. The probable origin of many, perhaps all, of the finer structured, or shorter wave-length billow clouds, is convection through unstable air beneath into air above of different velocity.

When the two superincumbent layers of air are still with reference to each other the ascending masses are roughly as broad as long, horizontally, as indicated by the roundish patches of cloud that often cap them, and usually occur in more or less distinct rows and patterns. When, however, they are flowing the one past the other the clouds are more or less drawn out in the direction of the motion of the upper air with reference to the lower, and if this motion is in the direction of an alignment of the clouds the individual small cumuli in each row parallel to the wind are so intermingled, or so it seems, as to become a uniform cloud wave or billow. This obviously is not the whole story of the genesis of the billow cloud, and frankly it may not be more than half right as far as it goes. At any rate this cloud still challenges us for much more observation, experiment and study than yet has been given to it, as do also cirro-cumuli waves, especially those that are crossed by numerous ripples.

Lenticular Cloud

Sometimes an isolated cloud, whether solitary or one of a group, has the general shape of a double convex lens, that is, thick in the middle and gradually thinning to nothing at the edge. This is the lenticular cloud, whatever else it may be, such as alto-cumulus, alto-stratus or what not. It may be formed by the arrest of an ascending small cumulus cloud at the base of a layer of stable air—a layer in which the decrease of temperature with increase of height is so slow that ascending air, unless in great volume, can not deeply penetrate it. Here the cloud, arrested by the impenetrable ceiling, spreads out much in the shape of a lens, circular or distorted as the circumstances, especially the direction and strength of the winds, may determine. Another abundant source of lenticular clouds is any large mountain peak, ridge or range. Here, in most cases, the crossing wind is thrown into great standing billows whose crests occasionally are capped with lenticular clouds as fixed in position as a waterfall, and made of as rapidly changing material—condensing on the windward side and evaporating to the leeward. Sometimes this lenticular cloud of obstruction or interference origin, and therefore cresting a stationary air wave, is low down, and again at considerable height, as determined largely by the amount and distribution of the humidity. This cloud also deserves more study than it yet has received.

Crest Cloud

When quite humid wind blows up and over a mountain a cloud forms, such as the famous *"TableCloth"* Cloth" on Table Mountain near Cape Town, that rests on the ridge and hangs down each side like a giant cloth, fringed on the leeward slope where it is pulled down in evaporating shreds by the descending air. This is the crest cloud. It is interesting and of some importance, and if baptized in Latin might eventually become a member of the aristocratic cloud family.

Riffle Cloud

Frequently, when the crest cloud is moderately heavy and the wind fairly strong one or more long cumulus billows hang fixed in the free air to the leeward of the mountain ridge and at about its height. These mark the upper portions of the great air waves caused by the interference of the mountain with the passing wind. Analogously produced waves, called riffles, occur in running water where it passes over an obstruction in the bed of the stream; hence the name "riffle cloud."

Banner Cloud

Any high mountain peak that steaks up like Cleopatra's Needle, and there are such—Mount Assiniboine in Canada, for instance—now then develops a cloud, in an otherwise clear atmosphere, that flutters out like a huge banner made fast to the peak's upper leeward side. This occurs, of course, only in air that is humid well-nigh to the point of saturation.

The cooling necessary to pass that critical cloud-producing temperature obviously may be owing in part to contact of the air with the cold walls of the peak, and certainly in part to the decrease of pressure in the eddy region on the sheltered side. It is not a familiar cloud nor one of much importance, but for all that it is both beautiful and interesting.

Boa Cloud

Sometimes a great river of cloud, such as may flow along a broad valley, is obstructed by an isolated mountain, a volcanic cone, perhaps, that stands in the midst of the stream and divides it right and left as would an island.

Thus the mountain becomes wrapt around by the flowing cloud, as by a giant feather boa. The old Romans called it, when it twined about Mount Etna, the *Serpe*, and rightly regarded it as a harbinger of rain.

Scarf Cloud

If one will watch carefully just above the top of a large growing cumulus he sometimes will have the pleasure of seeing a little cloud of silken sheen and bowed upward, over and free from the topmost head like a portion of a halo, emerge from the sky where before there was nothing to be seen. In a little while it probably will cover the head of the cumulus, and a bit later rest on its shoulders like a great silken scarf. It seems to form in a thin but quite humid layer of air that is cooled past the saturation point, and cloud thus formed, by the slight lift given to it by the ascending cumulus which it later adorns.

Some say that it is a harbinger of showers. Perhaps so; at any rate the towering cumulus that generates it certainly is.

MAMMATO-CUMULUS

If we ever saw the sky covered with huge grapes, or everywhere festooned, or a great mother cloud, as it were, plethoric with a thousand mammae, then, whether we knew it or not, we were looking at the interesting mammato-cumulus cloud, a frequent forerunner of the dread tornado. Not everything about the genesis of this cloud is known, but it appears to be the result of an upsidedown convection of some sort, perhaps induced by the flow of a thin cloudy layer of air over other air beneath that is sufficiently warm near its upper limit to permit of short-range interchanges between the two.

Tornado Cloud

The tornado cloud is a violently rotating, hollow, tapering cloud column 100 to 1000 feet in diameter, as a rule, extending from the surface of the earth, where it is smallest, up into the heavy storm clouds overhead. The cooling that produces this cloud is caused by the expansion of the air incident to the decrease of pressure within the column due to its rapid rotation. The cloud column of the waterspout is produced in the same way. This is a thin cirro-stratus-like cloud of rare occurrence and unknown origin, richly iridescent and about three times as high as ordinary cirrus, that is, around 16 miles.

Noctilucent Cloud

It was stated in the description of the cirrus cloud that usually it is the highest of all save two that are rarely seen. The higher of these, in turn, is the noctilucent cloud, a thin cloud-like patch of something, roughly 50 miles above the surface. Not all the clouds of this great height are lucent, but the first ones observed were, and the name "noctilucent" then given to them may as well be applied alike to all until we know something—now we know nothing-about the substance of which they are composed and how they are formed.

Chapter 6

Weather Measurements

As explained in the previous chapter, and as we all know without being told, it is our perceptions of the weather that both enable and compel every one to take a personal interest in it. And yet we can not go very far with perceptions alone in comparing the weather of one day with that of another, or the weather of one place with that of any other place. With respect to temperature, for instance, we may use such comparative expressions as cold, cool, comfortable, warm, hot; and that is about all. And even these few gradations are uncertain, for when the air is cool to one person it may be cold another; nor can either say with any assurance when it ceases to be cool and begins to be cold. And the situation is worse for most of the other weather elements. In fact, for some of them, such as atmospheric pressure and electrical state, we have no distinct perceptions at all. However, every weather element affects inanimate objects and in such manner that measurements of almost any desired degree of refinement are easily made and recorded, measurements, or instrumental readings, that are of great and increasing importance in the industries, that are the data from which weather forecasts are deduced, and that at the end of a few years' accumulation give us reliable knowledge of the climate of the place at which they were obtained.

EARLY BEGINNINGS OF WEATHER MEASUREMENT

No measurements of this kind were made in ancient times, except crude estimates here and there of the direction of the wind, and equally crude measurements of the amount of rainfall as indicated by the catch in certain exposed vessels. Both those very obvious and simple measurements have been made from time to time and at one place or another during the last 2000 years, and perhaps much longer. But no other weather element was measured—there was no instrument to measure it with—until less than 300 years ago; not, indeed, until, grown-ups, retaining and respecting to the inquisitiveness of the child, began to put questions direct to Nature instead of turning to Aristotle or supporting wild fancies by convenient interpretations of passages from the Bible. Instrumental meteorology was born at Padua, Italy, in the first half of the 17th century, the child of the thermometer, invented by Galileo, and of the barometer, invented by Torricelli. Since that time it has grown mightily, with many new features added and refinements acquired, but always its chief dependence has been on its parents, the thermometer and the barometer.

TEMPERATURE

Every one has a temperature sense, at least enough to distinguish between hot and cold, and he will readily agree that a hot object has a different temperature from that of a cold one, but even so he can not, without special instruction, know just what temperature is, or make any distinction between temperature and heat. Indeed the wisest in Nature's ways were themselves a long time in finding this out. All the Solons and Solomons down to near the middle of the 18th century appear to have accepted the existence of heat without attempting to determine its nature, or even to speculate about it.

Then, with the advent of the private laboratory and the eagerness to understand natural phenomena, came philosophizing and speculation, and hear came to be regarded as a sort of material substance, but one without weight, an imponderable, since by experiment objects were found to weigh just as much when cold as when hot. Near the close of the century (18th) it was found that much heat can be added to ice at its melting point without changing its temperature, from which it was inferred that heat and temperature are radically different things. At about the same time it was found that an indefinite amount of heat can be gotten from a couple of objects just by rubbing them together; and finally, near the middle of the 19th century, it was found that there is a definite relation between of work and quantity of heat.

Thus, tediously and laboriously, we learned that heat is not a substance at all, but a something interchangeable with work, or as we now call it, a form of energy—energy being capacity for doing work. But this doesn't yet tell us what temperature is. Perhaps we might regard temperature as a measure or index of the concentration of heat inasmuch as the same amount of heat, as near as we can judge it, makes a cup of water much hotter than a pot of water, provided neither is brought to the boiling point. This is easily understood, but like lots of other simple definitions and explanations it is all wrong, as Nature proves by citing an endless number of facts like these: The same amount of heat that would just melt a pound of ice, making no change whatever in the temperature, would heat a pound of ice water to a scalding temperature, 176° F., roughly, or seven and three-quarter pounds of ice-cold lead up to its melting point.

Evidently, then temperature is not a measure of the concentration of heat. But there is another experiment that helps us to a definition of temperature that appears to be satisfactory. The experiment is this: When an object that is decidedly cold to the touch is put in intimate contact with one that feels distinctly warm, and both are protected from contact with other objects, the cold one always gets warmer and the warmer one cooler. From this we conclude that temperature is that thermal state of an object that enables it to communicate heat to other objects. We may assume that there is a flow of heat from each object to the other, whatever their temperatures, as there certainly is when this exchange is by means of radiation, but in all cases the net result is a gain of heat by the colder body and a loss by the warmer.

Thermometers

It has been found by experiment that all objects change in volume with change in temperature, and this has furnished our most convenient and common method of comparing and labeling temperatures. A small bore glass tube, for instance, with a bulb at one end is filled to the desired point on the stem with mercury, or other fluid which will expand faster than glass, and sealed off, and the whole made into a thermometer is follows: First, the bulb and stem are placed in a mixture of pure ice and water, which always has the same temperature, and the end of the mercury column marked on the glass tube.

This point is labeled 0 (zero) if we are making a centigrade thermometer, or 32 if it is to be a Fahreheit thermometer. Next, the bulb and stem are immersed in a bath of live steam over boiling water and the new position of the end of the mercury column marked. The atmospheric pressure is noted at the same time because the temperature of the steam varies with this pressure. If the pressure is just right (if not right the proper correction can be made) the new mark on the stem is labeled 100 or 212 according as the thermometer is to be centigrade or Fahrenheit. Finally, if the bore of the stem is uniform, the distance between these two marks is subdivided as finely as desired and as many as necessary of the division marks appropriately labeled. Also similar divisions may extend beyond each of the initial end marks.

This construction of a thermometer is based on the fact that the volumes of the glass vessel and the mercury both increase almost uniformly with increase of temperature, but unequally—the mercury fastest. There are, of course, many refinements not indicated here in the manufacture of good thermometers, but the principle remains the same. Of course some other fluid than mercury may be used, and is used, for measuring temperatures below -40° F., the temperature at which mercury freezes.

Maximum Thermometers

When the tube of the thermometer is sharply constricted a short range, generally near the bulb, as in the case of the well-known clinical thermometer, the mercury passes by that place on increase of temperature due to the force of expansion, but does not pass it, the column promptly pulling in two, on decrease of temperature. It can, however, be thrown back by a proper swing of the instrument. This thermometer therefore can be, and is, used to mark the highest temperature to which it was exposed since it was last adjusted. It is the standard instrument used for obtaining the day-to-day maximum temperature of the air.

Minimum Thermometers

If alcohol is used as the thermometer fluid, and a miniature glass dumb-bell, colored so as to be more easily seen, is placed within the fluid in the bore of the stem, and the whole mounted horizontally, then with increase of temperature the alcohol flows along the tube past the dumb-bell, which fits loosely, and without disturbance to it. On the other hand, when the column contracts the surface film at its outer end drags the dumb-bell along with it. Hence this thermometer, much used in meteorological work, gives the lowest temperature, as indicated by the outer end of the sliding index, since its last adjustment—an operation that consists only in holding the thermometer sufficiently inclined to let the index drop into contact with the surface of the fluid column. The "empty" portion of the stem should contain air so as to reduce the trouble due to a shift of a portion of the alcohol to the outer end of the stem by evaporation and condensation.

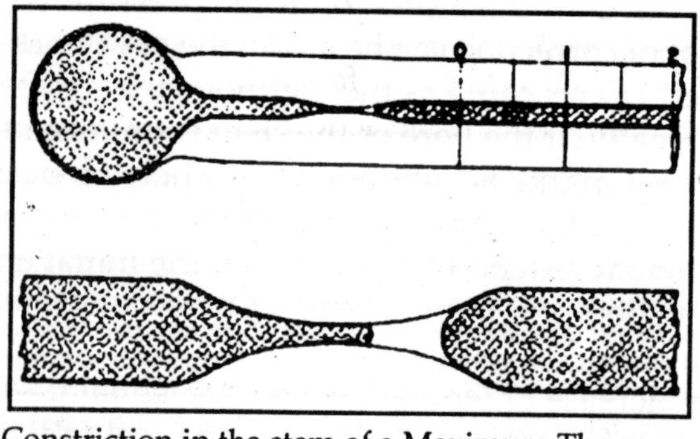

Fig. Constriction in the stem of a Maximum Thermometer.

Telethermoscope

Often while indoors we need to know what the temperature is outdoors, and therefore have devised instruments that give us that information without the trouble of going to the roof, or wherever the outdoor thermometer may be. An excellent instrument of this kind is based on the fact that the electrical resistance of a metal wire (pure nickel is excellent) varies with temperature. We therefore can expose such a wire, properly protected from injury, to the free air, and connect it with a small electric battery and a current indicator on our desk, if we like.

We then can vary an adjustable resistance in the circuit until it is equal the current indicator, to the resistance at that moment of the thermometer wire. If also the action that adjusts the balancing resistance correspondingly moves a pointer, a result easily secured, we can, of course, mark each position of that pointer with the then temperature of the exposed wire. After this we have only to close the circuit (open except when an observation is being made) and turn the pointer to the position of balance, a matter of only a few seconds, to read on our desk, and at any time, the temperature of the air outdoors, or of any thing else to which the thermometer is exposed.

This instrument is extensively used in meteorological offices. It enables the observer to answer frequent requests for the current temperature without leaving his desk to the neglect of other duties.

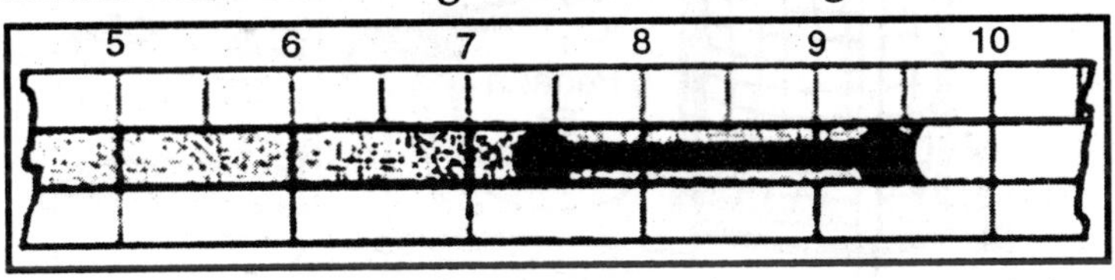

Fig. Dumb-bell in the Stem of a Minimum Thermometer.

Thermograph

Reading a thermometer every few minutes day and night, month in and month out, is wholly impracticable. Yet a continuous record of the temperature is desirable, for many significant changes occur at brief intervals.

Hence thermographs have been devised that keep an absolutely continuous record of the temperature of the air. The basic element of the most extensively used instrument of this kind. Consists of a thin-walled, flattened, metal tube; curved along the flat sides, completely filled with alcohol and tightly closed; in short, a Bourdon tube of the kind that is widely used in measuring gas and liquid pressures. With increase of temperature the enclosed liquid, expanding faster than the metal, tends to straighten the tube; and with decrease of temperature curves it more.

Hence by rigidly fastening one end of it to a fixed object, and a pen to the other or free end, we have directly, or through a multiplying linkage, a means of obtaining a continuous record of the temperature on a properly graduated moving sheet of paper.

Telethermographs

Obviously the recording device of a thermograph can be at one place and the thermal element at another, with either electrical or mechanical connection. Both kinds of instruments are in use, but they are not numerous since the need for them is relatively slight.

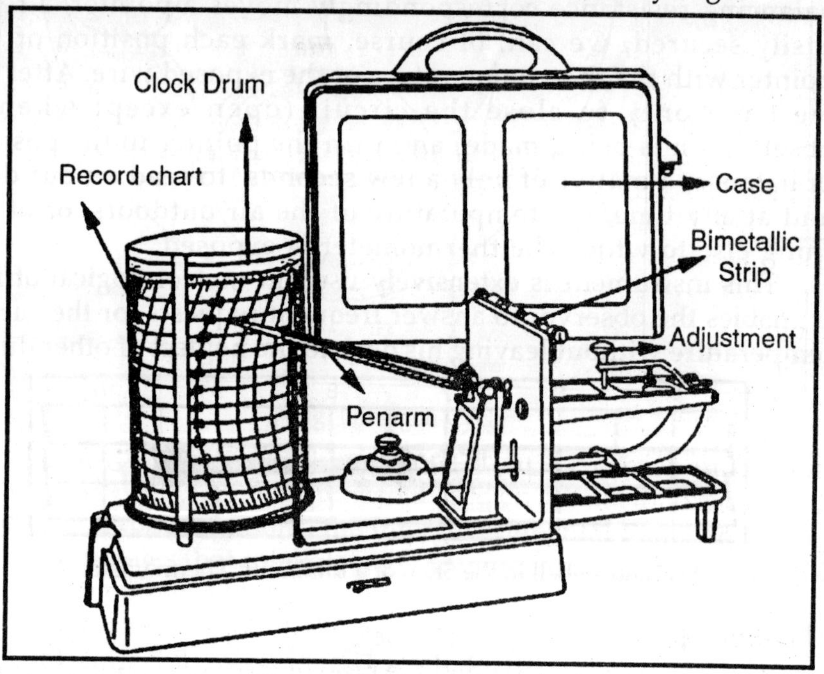

Fig. Thermograph.

Thermometer Shelter

To obtain the correct temperature of the air outdoors it is necessary that the thermometer of whatever kind be protected from sunshine, both direct and reflected, since either, and especially the former, would make it read much too high; also that it be protected from rain and snow as they would make it register too low.

At the same time it must be exposed to free and easy circulation of the air. These ends are sufficiently well obtained by housing the thermometer in a double roofed shelter with louvred sides, all painted white. This shelter does not secure entire freedom from errors, especially when the sun is shining on it, but they seldom are greater than a small fraction of a degree and there- fore may be neglected.

A much more important matter is the location of the shelter and its height above the surface since, in extreme cases, a difference of a few feet in height above the surface means a difference of as many degrees in the temperature of the air. On still, clear days the temperature often decreases very rapidly through the first 10 to 20 feet above a flat barren surface, and increases quite as rapidly during calm clear nights.

Hence it is highly desirable to have the thermometer shelter on a grass plot, since that does not heat up greatly, and on a slight elevation to avoid undue accumulation of cold air, and, finally, at a standard height above ground. In fact the shelter and its installation are quite as important in obtaining the true temperature of the air as is the accuracy of the thermometer.

Katathermometer

A good deal has been said and written about this kind of thermometer which, on exposure to the air, indoors or out, gives readings more or less indicative of the degree of comfort or discomfort a normal healthy person would experience in that air at that particular time. Essentially it is only a pair of thermometers, one of which is kept dry and the other (bulb and a portion of the stem) supplied with a drippingly (but not dripping wet jacket of muslin or other suitable material.

If both are allowed to cool simultaneously from some fixed temperature, a few degrees higher than that of the body, to some other fixed temperature, a few degrees below that of the body, it will be found that their rates of cooling, and the ratio of these rates, one to the other, depend on the temperature, humidity and movement of the air, and exposure to or protection from sunshine, that is, on the very things that, so far as the weather is concerned, contribute most to our comfort or discomfort.

Hence these readings can be, and they have been, coordinated roughly with our feelings. However, this device is not much used. Perhaps no one needs an instrument to tell him how he feels; while in the process of air conditioning, whether for manufacturing requirements or physical comfort, it is much easier, and generally much better, to measure and control the actual temperature and humidity of the air supplied.

PRESSURE

The pressure of the atmosphere, its push per unit area, or weight of a vertical column of it of, say, one square inch cross section where the pressure is measured, extending from that level to its outer limit, is one of the most important of all meteorological measurements, especially as an aid, when known at many places, to weather forecasting. And yet our senses do not make us aware of its magnitude or of its changes.

Mercurial Barometer

The action of an ordinary suction pump, the drinking of cider through straw (they didn't have soda water at the time we are talking about), and a lot of things of like kind, used to b explained by the learned as due to Nature's abhorrence of a vacuum.

Galileo is said to have remarked, ironically, it must have been, that in the case of water Nature didn't abhor a vacuum beyond about 30 feet, that being the limit to which water could be pulled up with a suction pump. Torricelli in experimenting on this problem took a long glass tube, closed at one end and open at the other, filled it with mercury and then stood it upside down with its lower end dipping into a basin of the same substance.

The mercury in the tube dropped until it stood about 30 inches above that in the basin, a height that is the same fraction of that of which water can be sucked as its density is of that of mercury. This showed that Nature abhorred a vacuum only to the extent of the weight of the air then above the place of abhorrence.

At sea level this generally is around 14 pounds per horizontal square inch. It decreases with increase of height by the weight of the air left below, a fact that affords a ready means of determining height without directly measuring them. It also generally varies with the kind of weather and strength of the winds.

The simple tube and basin of Torricelli are the essential elements of the mercurial barometer or measurer of the weight, or pressure, of the air by means of a balancing column of mercury. Ordinarily they are provided with a convenient supporting frame, scales, thermometers and adjustment screws; but, simple or complex, the purpose is to measure the pressure of the air at the level of the basin surface in terms of the height of a balancing column of mercury.

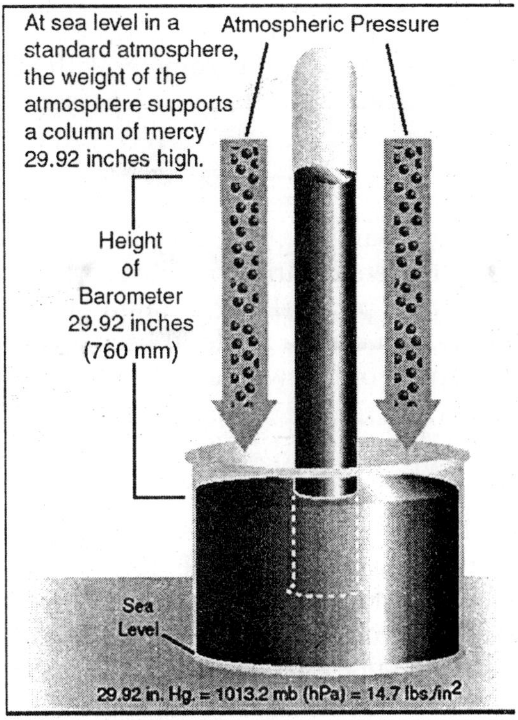

Fig. Mercurial Barometer

Aneroid Barometer

The pressure element of the aneroid barometer is a flexible metal shell, fully exhausted and sealed, with top and bottom held apart by means of a suitable spring. With change of atmospheric pressure the shell correspondingly contracts or expands, as the case may be, which movement commonly is translated into the travel of a pointer over a dial marked in terms of the corresponding readings of a mercurial barometer.

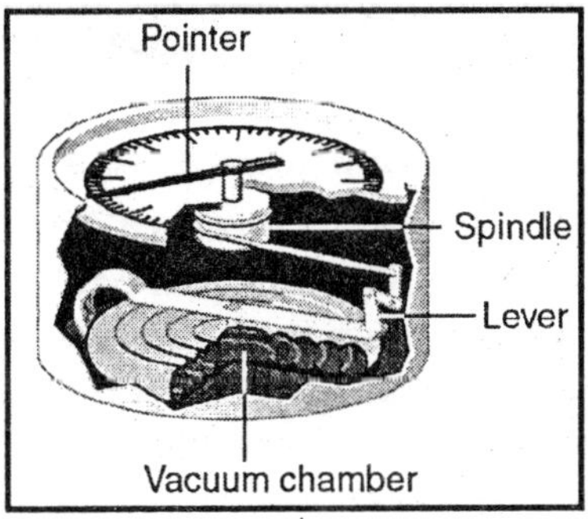

Fig. Aneroid Barometer

Altimeters

An altimeter is only an aneroid barometer with a scale in terms of heights instead of pressures—decrease of pressure indicating an increase of height.

Barograph

The most common barograph, is only an aneroid barometer so constructed that the traversing arm writes an continuous record on a uniformly moving and properly graduated sheet of paper. The mercurial barometer also can be made to keep a continuous record, graphic or photographic, of its own height.

WIND

Wind Velocity

The velocity of the wind can be measured more or less accurately by many different means. However, the device commonly used for this purpose is the Robinson cup anemometer, the best form of which consists of three or four equally spaced hemispherical cups that open horizontally, face in the same direction around their common path and are attached through short arms to a vertical shaft which, through proper gearing, momentarily closes an electric circuit at the end of every so many revolutions.

Fig. Barograph.

At each closing of the circuit a record is made, in the office or elsewhere, on a uniformly moving sheet of paper. The instrument may be so geared that a contact occurs for each mile, approximately, travel of the wind, or other distance as desired. This value, together with the known rate of movement of the record paper, gives the average speed of the wind over whatever appreciable interval, and at whatever time, one may wish to know it.

Quick Acting Anemometers

As usually constructed the Robinson cup anemometer just described is too sluggish, has too much inertia, to respond promptly to changes of wind velocity and therefore is not adapted to the measurement of wind gusts and eddies. However, there are various more or less elaborate ways by which such measurements can be made.

A tube, for instance, can be kept facing the wind and the pressure inside registered through the resulting height of a balancing liquid column; or the cooling of an electrically heated wire, exposed to the wind, may be measured by its resistance and, by proper calibration, the wind velocity and its changes thus determined quite accurately. Various other methods of making a quick acting anemometer readily suggest themselves, but, although greatly needed, none has yet come into use that at once is simple, sturdy, accurate, continuously recording and inexpensive.

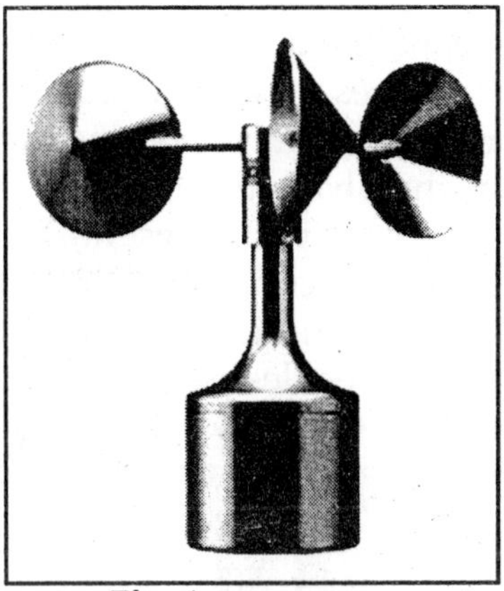

Fig. Anemometer.

Wind Direction.

The direction of the wind is best and most easily obtained by the use of a wind vane, a horizontal arrow with a vertically broad tail so mounted on an upright rod as to turn easily with the wind. The supporting rod may turn with the vane, and may carry a pointer at its lower end moving just beneath a dial on the ceiling of a room or portico beneath, from which one may see the pointing of the vane and thus know the direction of the wind without going out of doors.

This ingenious device was one of the luxuries of Roman villas 2000 years ago, and also one of Jefferson's many conveniences at Monticello. However, as commonly used at meteorological stations, the turning rod carries certain electrical contacts by which registration to eight directions is secured on a moving sheet of paper at any desired location. It also is practicable to obtain a continuous record of wind direction by means of a vertically moving pen over a sheet of paper on a cylindrical drum carried by the vane rod.

Recently the vane has been rendered much more sensitive by making the tail blade relatively short and of stream-line shape —blunt and rounded in front and tapering to a vertical edge in the rear— instead of long and flaring, with the edge in front.

HUMIDITY

There are many ideas about humidity, mostly vague and erroneous. The correct ideas are but few. The humidity, or water vapour content, of the air is expressed in various ways according to the purpose in view.

Absolute Humidity

This expression means the actual mass of water vapour present per unit volume; and it makes no difference whether the other gases of the air are present or not. They have nothing to do with the humidity. We often say, too, that the absolute humidity is the actual pressure per unit exerted by the water vapour present. This is not the weight of the water vapour in a vertical column of unit cross section, but just that fraction of the total atmospheric pressure, at the place in question, that the number of water molecules per unit volume at that place is of the total number of molecules of all kinds in that unit volume.

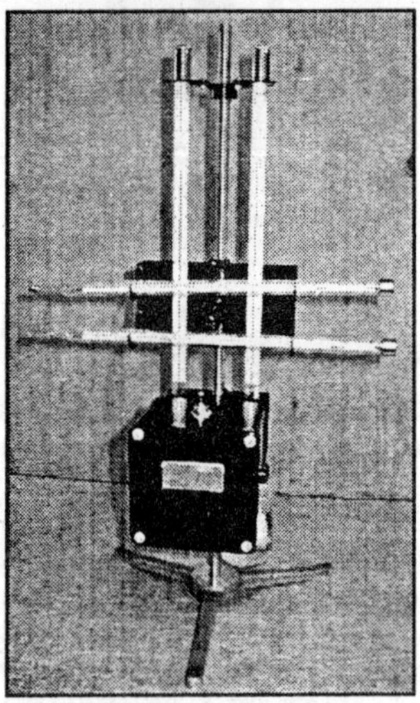

Fig. Psychrometer.

At any given temperature this vapour pressure per unit area is directly proportional to the mass of water vapour per unit volume. Hence, if we know what we are doing, we may express humidity in terms of either mass or pressure. Absolute humidity seldom is measured directly in meteorological observations.

Relative Humidity

Relative humidity is that fraction, or percentage, which the actual water vapour present per unit volume is of the total amount that unit volume could contain at the same temperature, and in the presence of a flat surface of pure water. Commonly one says, "per unit volume of air," "total amount the air could contain," and the like, but that is all wrong. The water vapour is not contained by the air, it is a part of the air; neither do the other gases of the atmosphere soak up water vapour, as a sponge takes up liquid water. In fact, they do not appreciably affect the amount of water vapour per given volume necessary to produce saturation, or any particular fraction of saturation. That is determined by temperature alone. In meteorological work relative humidity is determined indirectly and by reference to tables or graphs previously established by elaborate experiments under known conditions.

Psychrometer

The psychrometer or instrument generally used for determining the humidity, consists of a pair of thermometers, one dry, the other wet (having a closely fiting jacket of wet muslin) at the time of observation and provide with some means of obtaining ample ventilation. The dry thermometer gives the current temperature of the air, while the wet one indicates the cooling due to evaporation, a quantity that depends on both the actual temperature of the air and its relative humidity, or, if we prefer, humidity deficit (difference between saturation and the current humidity), or dryness. These numerical relations have been determined empirically by numerous careful experiments and recorded in convenient tables, so that all we now havc to do is to read both the wet bulb and the dry bulb thermometers and then look up in printed tables the relative humidity, absolute humidity and dew point, or temperature at which the vapour actually present would produce saturation.

Hair Hygrograph

It has been found experimentally that at all ordinary temperatures of the atmosphere the human hair, when clean and oilless, responds but little to temperature changes but appreciably lengthens with any considerable increase of the relative humidity. Hence, although the rate of this gain in length to a given increase of the relative humidity decreases as saturation is approached, it is possible, with a small strand of hairs, acting through a suitable cam, to make a tracing pen so traverse a moving graduated sheet of paper as to produce a reliable record of the varying humidity.

CLOUDS

Amount

The amount of cloudiness, or portion of the sky covered with clouds, is a very important weather element, but ordinarily it is only crudely estimated in terms of tenths of the sky that appears to be covered. Usually, too, no distinction is made between very thin and very thick clouds, although they differ greatly in respect to the amount of sunlight that gets through them.

Height

The height of a cloud canopy, now an exceedingly important matter in connection with aviation, can be determined fairly closely by any one of several methods. Evidently all aviator can fly up to the cloud and note the reading of the altimeter when he gets there. Or, without going up oneself, one can set free a pilot balloon (a small rubber balloon), so inflated as to have a known rate of ascent, and note the interval from the time of launching to disappearance in the cloud. This interval multiplied by the rate of ascent gives the approximate height of the cloud above the observer.

Greater accuracy is obtained by following the balloon with two theodolites at known positions some distance apart. From the pointing of the two instruments at the instant the balloon disappeared the exact height of the cloud is readily computed. Evidently, too, the height of a cloud could be gotten fairly well in many cases, with a good range finder.

A simple way of getting this height at night is to throw a parallel beam of light, search light or "ceiling light," onto the cloud at a known angle of elevation and then sight the illuminated spot, noting also this angle of elevation, from a point at a known distance from the light, at its level and in the same vertical plane as the light and spot. The height of the cloud may then be read on the pointer, if the other values have been properly chosen, or in any case, easily computed. If the beam is turned up 45° then clearly the height of the cloud is equal to the distance from the light to that point on the ground which is directly beneath the illuminated spot.

Velocity

There are also several methods of measuring the velocity of a cloud. We can follow some recognizable feature with one or two theodolites and compute the velocity by triangulation. If it is an isolated cloud casting a shadow we may be able to get the velocity of the shadow, which, evidently, is the same as the velocity of the cloud casting it. Or we may watch through a fixed peep-hole the travel of a cloud image over a horizontal mirror—a nephoscope, as it is called — and compute the value desired from the obvious fact that the velocity of the cloud is to the speed of its image across the mirror as its height above the observer is to the height of the peep-hole above the mirror. This assumes that the height of the cloud is known. The nephoscope also, and always, gives the direction of travel of the cloud.

Kind of Cloud

The kind of cloud, as well as its amount, velocity and direction of travel, is noted in meteorological observations because, in connection with other things, it is significant of the coming weather, and because it is a climatic element, though one of quite secondary importance. And the kind can be recorded either by its recognized name, or by letter or number corresponding to a particular cloud picture.

The names that refer to the primary or fundamental forms are: *cirrus*, or curl cloud; *stratus*, or layer cloud; *cumulus*, or pile cloud; and *nimbus*, or rain cloud.

But these few names are not enough to go around, so we have also combination names; cirrostratus, cirro-cumulus, strato-cumulus and cumulo-nimbus; names with "alto" as prefix, signifying high, that is, high for that particular kind of cloud; alto-stratus, and alto-cumulus; and names beginning with " fracto" to indicate a ragged state of the cloud; fracto-stratus, fracto-cumulus, and fracto-nimbus; and, finally, names descriptive, usually, of appearance of a number of clouds that differ in some important respect from any of the above forms, such as the mammato-cumulus, a sort of a festooned cloud; the funnel cloud, characteristic of the tornado; the banner cloud, a cloud floating like a banner from a mountain peak; crest cloud, the cloud that envelops and stretches along a mountain ridge; and an indefinite number of still other names, generally self-explanatory.

PRECIPITATION

The amount of snowfall and rainfall, and the time, rapidity and frequency of its occurrence all are important matters. Systematic observations of the amount of rain caught in properly exposed open-mouthed vessels have been made, at least sporadically and in various countries, for many centuries. In Palestine, for instance, quantitative measurements were made in the first century A.D. Today they are made in great numbers in every progressive country, and the method of making them is everywhere, and always has been, essentially the same.

Fig. Rain Gage.

Rain Gage

Every rain gage, however much it may differ from others (and there are many kinds), is an open-mouthed vessel set upright where the amount of precipitation is not appreciably affected by the surroundings. To be at all accurate the mouth of the vessel must be horizontal and sharp lipped so that all the precipitation that hits the mouth, and none other, will be caught. To decrease the errors of measurement the catch may be funneled to a cylindrical vessel whose cross section is a known fraction, a tenth, say, of that of the exposed mouth. The rainfall then, in terms of the horizontal layer of water it would have produced if none had run off, or otherwise disappeared, is one tenth, or whatever the fraction may be, of the depth of the water in the smaller vessel.

A continuous record of the rainfall may be obtained by allowing the water, as it is caught, to run through a small pipe into a little rocking bucket, divided into two equal parts by a partition, and so adjusted that as soon as water representing one one-hundredth of an inch of rainfall, or any other predetermined amount, has been caught in one compartment it tips over, thus emptying that half and catching in the other, and so on as long as the rain lasts.

The bucket can be so connected in an electric circuit that each tip it makes, hence every hundredth, say, of an inch of rainfall, is automatically recorded on a moving sheet of paper. In this way the quantity, intensity and time of occurrence of each rain is permanently recorded. If desired the catch can be piped into a suitable vessel supported by a weighing device adapted to either occasional reading or continuous recording. Evidently, too, a record may be obtained on a moving sheet of paper by a pen attached to a float in the rain catch vessel. But whatever the kind of device used for measuring rainfall and however various the details, the fundamentals remain the same.

Snow Gage

When the precipitation is snow it is more difficult to obtain a true catch, owing to the tendency of the wind disturbances, induced by the vessel itself, to keep the snow from falling into it, or even to whisk from the gage a portion of that already captured. Also drifting snow may be caught and thus precipitations recorded when actually there were none at all—only shiftings of snow that previously had fallen. These troubles are reduced by placing a downward deflecting shunt around the gage at the level of the catchment mouth. The snow thus caught may be melted and the resulting water measured, or it may be weighed and the equivalent rainfall computed.

Snow Tube

Often it is desirable to know the mass or water equivalent of the snow at a particular time over a considerable mountain area, especially for computing the probable runoff available later on for power and irrigation. This information can be obtained by pushing a thin-walled, open-minded, metal tube vertically through the snow and thus picking up a column of the snow of known cross section. The lower end of the tube should be rather sharp lipped and slightly smaller in diameter than the main body to prevent packing. The difference in weight between tube plus snow core, and tube alone, gives the weight of a column of snow of the known cross section. This is readily reduced to equivalent water depth, or, for convenience, the scales may be graduated in terms of water depth and set to read zero when the tube is empty. By taking many such samples properly distributed the available runoff from a given reservoir of snow can be fairly closely determined.

Hail Gage

If the catch of a rain gage is immediately passed over a sloping section of wire gauze in such manner as to shunt the hail stones, when there are any, into a special compartment the volume, or weight, of the water they produce on melting is, of course, a measure of the hail fall. To insure a catch fairly representative of the fall, especially if it happens to be sparse, the mouth of the gage must be quite large. If the water that drains through the gauze is also caught, separately, the combination might be called a rain-hail gage.

Drosometer

This instrument that weighs the amount of dew that collects on a given area of any chosen material is very little used since the quantitative values thus obtained have practically no current use. However, the frequency of the occurrence of dew, month by month, or week by week, and its relative amounts—light, moderate, heavy—is an important climatic factor in respect to certain crops and their diseases. Grapes, for instance, may succeed admirably at one place, where the dews are few and light, and fail at another, less than a mile away, owing to fungus diseases that are serious where dews are frequent and heavy.

EVAPORATION

Evaporation occurs from growing leaves, damp soil and free water surfaces. However, measurements of evaporation generally are concerned with the latter, or free water surface, only. And even in this case it is not the evaporation that is measured but the net evaporation, that is, the difference between the total amount of water that left the surface under consideration as vapour, and the amount that in the same time, condensed onto it as water from the vapour in the adjacent space. This difference indeed may be either positive, evaporation greater than condensation; zero, evaporation and condensation equal; or negative, condensation greater than evaporation, according as the temperature of the water is greater than, equal to, or less than the dew point of the nearby air.

The common method of measuring net evaporation is to note the rate of fall, or rise, of the surface of water in a suitably exposed vessel, usually a cylinder 2 to 4 feet in diameter and a foot deep. Innumerable measurements of this kind have been made under various conditions, from which we know that evaporation varies with the temperature and salinity of the water, fresh water evaporating about 5% faster than sea water; condition of the surface (clean or foul) ; temperature and humidity of the air; area of the surface; and wind velocity. Nevertheless, no exact, useful and all-inclusive equation has been found for evaporation. Indeed, it is practically certain that no such equation can rigidly apply to natural bodies of water, such as lakes and ponds. Of course the amount of evaporation by a given body of water is proportional to the energy it absorbs from all sources during the time in question minus the energy it loses in every way except by evaporation. But this does not necessarily help much since each of the quantities of energy involved may be more difficult to determine than the evaporation directly.

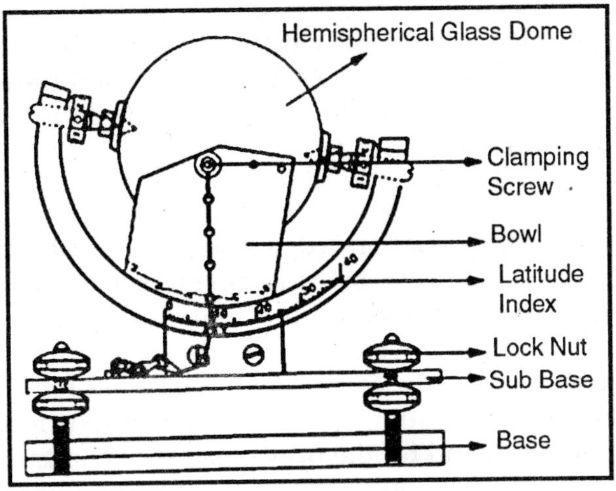

Fig. Sunshine Recorder.

INSOLATION

Fundamentally, sunshine is the most important of all meteorological and climatological elements, and it is measured in respect to different factors.

Duration

One measure of sunshine is just its duration, either in hours and minutes or as a percentage of the total possible, and the times of its occurrence. This can be done by using a glass sphere as a burning lens to sear a line, during sunshine, on a curved and properly ruled sheet of paper. The same measurements are made also with a vacuum-jacketed differential air thermometer, one bulb blackened, the other bright with a plug of mercury in the connecting stem, and so adjusted that while the sun is shining, and only then, the position of the mercury is such as to be in touch with a couple of sealed-in wires. This affords the means of obtaining, with an electric current, an automatic record of the duration, but not the intensity or quality, of the sunshine. Various other devices could be used to accomplish the same ends. Those outlined are only the two in widest use.

Intensity

The intensity of the sunshine is measured in terms of the amount of energy it delivers per unit of time, over a unit area that either directly faces the sun or is horizontal.

The radiation is caught by some form of a black body, or perfect absorber, that is protected from disturbances from other sources than the sun, and its intensity indicated by the rate of heating of the receiver by a thermometer, mercurial or electrical; by change of voltage of a thermo-couple, or otherwise. Also every pyrheliometer, as this instrument is called, is provided with a number of necessary and convenient accessories.

Quality

The quality of the radiation is determined both as to kind, or wave length, and relative intensity, by traversing its spectrum across a thin, narrow strip of blackened platinum which continuously records, automatically, its changes in electrical resistance, incident to variations of temperature; and, finally, correcting these values for the inequality of the dispersion or spreading out of the spectrum (the width of the strip being constant), and for the previous loss of energy by reflection and absorption within the instrument.

SKY LIGHT

The light of the sky is measured in terms of its brightness over various portions, colour, and polarization—a state caused by scattering and reflection, but not ordinarily perceptible to the eye.

EARTH RADIATION

Evidently the earth loses heat by radiation at rates that vary with the temperature and the nature of the radiating surfaces. Many efforts have been made to measure this radiation, but with indifferent success.

In fact, the measurements actually made usually are of the difference between the energy lost in a given time, per unit horizontal area of a black body, to the sky, and the energy simultaneously absorbed from the sky by that same area. If the temperature of the black body is known, as it may be, its total radiation is easily computed; and if the black body is so protected as to lose and gain heat by radiation alone, this total computed loss, minus the measured difference between loss and gain, gives the energy received per unit horizontal area from the sky.

The radiation from the surface of the earth probably can be

computed more reliably from its temperature than actually measured, since it loses heat nearly as does a black body.

ATMOSPHERIC DUST

The dust in the atmosphere may be classified in many different ways. One kind is the relatively gross particles. These are most abundant in the lower air, often 10,000 to 50,000 per cubic inch, especially in cities and over desert regions.

These affect our health and our industries and are largely responsible for the haze that limits the vision of the aviator. It therefore is important that the quantity of this dust be measured, and the things it consists of determined.

Both can be done by first humidifying the air, then drawing it rapidly into an expansion chamber through a small opening—a narrow slit is especially convenient—in such manner that it will impinge squarely against a flat object, such as a microscope cover glass, and finally examining with a suitable microscope the material thus caught sticking to the baffle surface owing to the water condensed on the particles incident to the cooling of the expanding humid air. Another method of catching the dust is to cause the air, in its current condition, to impinge gently against a surface made sticky with a very thin film of vaseline, or other suitable substance.

The material of this dust includes many kinds of rocks and minerals, soot, sea salt, pollen (in season), spores and every sort of vegetable fibre. Indeed one feels like wearing a dust mask, when he recalls the fact that at a single breath he may inhale a million sticks and stones of a hundred different kinds!

But this is not the whole story. There is another, though overlapping, kind of dust that consists of particles often far more numerous than those just discussed; most of them smaller and many of them liquid or even gaseous.

This "dust" consists of all those particles, solid, liquid or gaseous, upon which water vapour readily condenses as the temperature falls below the dew point—the condensation nuclei of fogs and clouds. It will not pass through a plug of raw cotton or similar material. The number of these particles, ranging from a few hundred to more than a million per cubic inch, may easily be estimated by saturating the air to be examined in a closed vessel, then giving it two or three sharp, or fog-producing, expansions and counting the number of droplets caught per unit area on the lower of two horizontal surfaces a known short distance apart.

For convenience the surface on which the droplets fall may be of polished silver lightly ruled in squares of known area. There generally are other conveniences also, microscope, lamp, pump, filters, et cetera, but the above are the essentials of the Aitken dust counter, named after John Aitken, the first to give much attention to this subject.

This sort of "dust," essential to the formation of every fog and cloud, has been under investigation now for more than 50 years. We know that nonhygroscopic particles of whatever origin are no part of it. We know, too, that particles of sea salt, gotten into the air by evaporating spray, are a large portion of it; and that many other substances, even certain gases, act in this capacity. Nevertheless, the problem of the condensation nuclei still is but imperfectly solved.

VLSIBILITY

The degree of transparency of the air is very important to the aviator. It means, as the case may be, good seeing and knowing all the time where he is, or bad seeing and getting lost.

Poor seeing generally is due chiefly to glare, caused by the reflection and scattering of light by particles in the line of sight. This glare commonly is reduced by a passing cloud. Also it can be reduced by looking through yellow to reddish glasses since light of this colour is not nearly so much scattered by dust as is the blue, for instance. If the poor seeing is owing to actual fog it is but little improved by screens of kind since fog scatters all colors about equally. The usual method of measuring visibility is the crude one of noting the maximum distance at which houses, trees, et cetera, can be recognized—seen well enough to know what they are.

ELECTRICAL STATE

The electrical of the atmosphere may be measured, but seldom is at meteorological stations, since it has but little weather significance. The factors most frequently measured are the difference in voltage between two horizontal levels, a known distance apart; the number of ions or electrified particles, both positive and negative, per unit volume; and the electrical conductivity.

FREE AIR AND UPPER AIR MEASUREMENTS

All the foregoing concerns measurements made near the surface of the earth. If similar measurements are to be made at considerable heights then evidently we must take the necessary instruments there ourselves in aeroplanes or balloons, or send up self-registering, or automatic wireless reporting instruments with free balloons of adequate size. All these means are in abundant use, and the basic instruments employed are fundamentally the same as those already described. They embody many ingenious details, to be sure, but like most other details, they may not be tomorrow what they are today, and therefore will not be further considered.

Chapter 7

Thunderstorms

People vary greatly in their reactions to thunderstorms. Some few take to feather beds in the fear that Thor will personally seek them out for punishment and destruction (though a feather bed actually avails against a lightning bolt about as effectively as a garden gate stops a railroad train). Other people, perhaps equally foolish, rush to hilltops in the fond faith that Divine Providence will somehow ward any wandering electric bolt away from their exposed heads. Actually a thunderstorm is a beautiful and gigantic experiment in physical science, doubly interesting to those who know something of its causes and its mechanism, wildly beautiful to all men conscious of beauty in Nature.

On what earthly mountains does a sunset glow compare with its pure rose radiance on the white cloud-cliffs of a towering and detached thunderstorm? What ground outlook compares with the spectacle of a distant towering cumulus in the clear night sky, flickeringly lit and veined by an infinite and never-ceasing variety of soundless lightnings? A great, wide-spreading, fast- traveling line squall, in its most violent and magnificent form, may be a towering range of such vapour-mountains, extending across the country for hundreds of miles and advancing in line like the fire-spitting front of some colossal and destructive army.

THOR'S APPROACH

Beautiful though thunderclouds are at a distance, either from the ground or from the air, on closer approach their still, turreted, snow-white serenity changes abruptly to a dark gray void of whirling winds, blinding firebolts, and torrential rain. If we (unwisely) fly into a thunderstorm in an airplane, this transition is very abrupt — one moment we are admiring the infinite overhang of a thousand cloud cliffs all gleaming with every gradation of sunlight and shadow; the next we are plunged into turbulent blue-gray obscurity, split by recurring yellowish flashes, deluged with rain, and disrupted by updrafts and air bumps of unbelievable ferocity. On the ground or the water, as the storm moves towards us at perhaps twenty or thirty miles an hour, the transition is slower and more obscure, yet spectacular and interesting enough.

It does not matter particularly whether we happen to be in the path of a single large summer 'air-mass' thunderstorm, or whether we are caught in the five-hundred-mile sweep of a spring or summer 'frontal' storm or line squall — the general effects will be the same, except that the line squall may be more violent, and its squall wind, temperature drop, and pressure rise will be more pronounced and permanent. In either case, reduced to typical form, the preceding summer weather is usually sultry and oppressive, both temperature and dewpoint being high. The sky is generally clear, but somewhat obscured by haze and cirriform clouds out of the west. The wind is southerly, and perhaps light or occasionally lacking altogether with a single thunderstorm in the offing, but likely to be strong and gusty if a line squall (cold front) is bearing down from the west or northwest.

If detached cumulus clouds are watched closely, they can often be seen to grow vertically upwards, sometimes at surpris-ing speed. Towards the middle of the afternoon a wall of thunderheads may be glimpsed, beyond obscuring haze and lower clouds (and are often plainly to be seen from an airplane at high altitude), rising slowly out of the west. But the thunderheads themselves often stream out into cirrus-like nebulosity at higher levels, and this, combined with ground haze, may hide them from view.

In any case, the westering sun is soon blotted out, and the sky to west and northwest slowly darkens, at least in its lower regions, to an ominous 'blue-black colour. Thunder begins to be heard perhaps half an hour to an hour before the storm strikes, and becomes louder and more frequent by the minute. The wind, which has perhaps died down altogether, rises and begins to blow directly towards the storm. Perhaps fifteen minutes before the squall strikes, the 'squall head' or roll cloud, whitish against the black clouds above and behind it, may be seen advancing in the shape of a long and irregular roll across the front of the storm — and this roll sometimes visibly revolves like a sort of horizontal tornado.

Perhaps fifteen minutes or more before the storm strikes, also, distinct lightning bolts come into view, most of them striking directly downward into the ground from the bottom and front of the black clouds just behind the roll cloud. We can check the narrowing distance away of these fiery ground-strikes, of course, by noting the flash-thunder time interval with a stop watch — the lag between sight and sound (as sound travels about eleven hundred feet per second in air) is about five seconds per mile. After the lightning bolts approach within a mile or two, it is only a moment (in a severe storm) before they are striking all around us, and in this 'dry' front of the storm, before the rain begins, the lightning danger is greatest.

A few minutes after the roll cloud passes overhead, the squall wind sweeps out of the west or northwest, usually at something like thirty or forty miles an hour, but occasionally as a flattening blast of sixty or more.

Along with the squall, or shortly after its onset, comes a wall of spattering, large-drop rain that may descend in torrents, and may be accompanied by lashing hailstones. In a detached 'air-mass' storm the squall wind usually begins to moderate after a few minutes, and may have died away altogether in a half-hour or so. The rain is less heavy, and the lightning less frequent and less intense, in the rearward and less active parts of the storm. But after a line squall the high westerly wind may continue or even increase.

THUNDERSTORMS OCCUR

Thunderstorms occur most frequently in the warm, moist, and unstable air masses within the tropics, where (in the rainy season) they are a daily afternoon occurrence, and sometimes so regular that they serve as a convenient time indicator for human affairs. In arctic regions, where the air is cold and comparatively dry, thunderstorms may not occur oftener than once in several years. Temperate regions experience their share of storms in the late spring and summer. No part of the United States is free from thunderstorm activity, but different regions vary widely in thunderstorm frequency.

In winter the centre of electrical activity is over the lower Mississippi Valley; in summer there are two far more active centers, one over Florida and the eastern Gulf coast, the other over northern New Mexico. But these centers are only relative, of course, for many parts of the nation see nearly as many storms. And in some regions, such as the Central, North Central, and East Central States, thunderstorms are likely to be mostly of the frontal type, and hence more severe and destructive.

Thunderstorms are classed, according to their apparent severity, as 'mild,' involving few if any ground-strikes and relatively light winds; 'moderate,' involving fairly frequent groundstrikes, moderate to heavy rain, and moderate to strong winds; and 'severe,' attended by nearly incessant lightning, torrential rain, and severe squalls. Temperate-zone thunderstorms are also classed, according to the mechanism of their formation, as 'airmass' and 'frontal.' The older names for these same classes of storms, before air-mass-and-front analysis came into wide use, were 'convective' and 'cyclonic.'

Air-mass thunderstorms occur entirely within one air mass, and without aid from any frontal action between air masses. They arise where the characteristics of the air mass, aided by wind and topography, favour them. They are most frequent in the United States in the warm, moist, and conditionally unstable air masses of tropical maritime origin. But moist polar air masses advancing southward rapidly, and hence remaining relatively cold in their upper levels, can occasionally be unstable enough to produce scattered air-mass thunderstorms.

The atmospheric instability that produces air-mass thunderstorms and keeps them active for hours over land areas in daytime is usually 'triggered off' by intense solar heating at the surface, which steepens the vertical temperature gradient; or perhaps by the importation of cold air aloft; or by the rising currents produced by upward deflection of horizontal winds (together with perpendicular solar heating) on the slopes of hills or mountains. Thus air-mass storms are frequent along the Gulf coast, where the sea air strikes the hot and slightly rising land; and thunderstorms of any type are more frequent over mountains than over flat country or at sea.

Frontal thunderstorms may occur along a warm front, where even gradual lifting on the warm-front slope is sufficient to trigger off the instability of very warm and very moist tropical air. They may also occur along the wave-like oscillations of a more or less stationary front that is, for the moment, alternately warm and cold. They may even arise, rather mysteriously for unaccustomed forecasters, out of upper fronts that give little or no other indication of their motivating presence. But the great majority of frontal thunderstorms occur either directly in front of, or actually ranged along, a vigorous cold front marked on the weather map by large temperature contrasts, a deep pressure trough, and strong shifting winds (that is to say, by a fairly strong pressure field having isobars sharply bent into V-shape where they cross the front). This whole formation is, of course, nothing more or less than a line squall.

Thunderstorms of all types are most frequent, on land, during the middle of the afternoon — that is, during the hours when solar heating at the surface reaches its maximum and the lapse rate (vertical temperature gradient) is consequently steepest. As the average life of a thunderstorm is six or seven hours, however, thunderstorms formed in the afternoon often persist into the early evening. Over the ocean, where the equable water warms up by only a degree or so during the day, and cools off by even less at night, the thunderstorm maximum comes during the late night and early morning hours, when radiational or airmass-transport cooling aloft is greatest, and, again, when the total lapse rate is steepest.

As the mechanics of frontal thunderstorms depend not so much on surface heating or upper cooling as on the violent upthrusting of warm and moist air masses by an undercutting wedge of cold air, frontal storms may occur at any time of the day or night. But even the frontal storms are most likely, and most severe, during the middle of the afternoon on land or during the early morning hours at sea. Plenty of active summer cold fronts, for example, do nothing much over land during the morning hours, yet develop into roaring line squalls by mid-afternoon.

FORMATION OF A THUNDERSTORM

What are the aerological conditions that favour the formation of a thunderstorm? First, there is the all-important factor of atmospheric instability, evidenced by a steep or steepening lapse rate (-3° F/1000 ft or steeper, or about -6° C/km or steeper). This steep temperature gradient is perhaps being increased by air-mass transfer below or aloft — lower warm winds from the south and southeast and/or upper cold winds from the north and northwest — and above all, by excessive solar heating at the surface.

In addition to the steep temperature gradient, high relative humidity at all levels (meaning that clouds will be formed by very little rise of air) also favors the formation of a storm. Finally the 'ice level,' or altitude of freezing temperature (+32° F. or 0° C.), should be located at considerable elevation, as it always is over the tropics. These upper-air characteristics, of course, can be fully known in advance only by means of airplane or balloon soundings, though they can sometimes be sensed by the rapid (and occasionally prodigious) vertical growth — the spectacular upward mushrooming — of detached cumulus clouds during the late morning.

There are also some surface thunderstorm portents not without value — high temperatures, say 80, 90, or 100° F. by day and 70° or 80° by night; high surface specific humidity as indicated by high dewpoint, say more than 6o° or 70°; and steadily decreasing surface pressure during the late morning and early after. noon. Another useful early-afternoon indication is increasing 'crash' static from a radio set — actually radio waves from lightning flashes beyond the horizon. Finally there are the definite evidences of frontal activity and advance towards one's position to be obtained from some sort of weather map — even a simple regional map that anyone can draw from airways radio weather reports in a few minutes' time.

One or more of the indications just listed can of course be present without resulting in thunderstorms, and thunderstorms can occur when one or more of them are absent; but in general the presence of most of them makes thunderstorms likely, and the agreement of all of them makes thunderstorms certain.

Given these favorable conditions, how does an air mass thunderstorm actually form and grow? The beginning of a thunderstorm is similar to the formation of an ordinary cumulus cloud or an ordinary shower. But in the case of the thunderstorm the air is more moist and more unstable, and this moist instability extends to greater heights. Hence the updrafts are more violent, and the cloudbuilding is more rapid and extensive.

As in ordinary cumulus-building, the rising currents of a thunderstorm cause the warm, moist air to cool by expansion to its dewpoint, forming minute liquid droplets (on commonly present condensation nuclei), perhaps 1/2500 inch in diameter, which droplets appear as the white or gray opacity of the cloud; and this condensation of water vapour, in a vicious cycle, releases latent heat which helps to drive and intensify the rising currents, which condense further moisture, and so on. The result is that under favorable conditions the thunderstorm grows fast and far — perhaps up to altitudes of six or seven miles.

Just as in the formation of all rain, so, while the minute cloud droplets of a thundercloud are carried up in the rising currents, there is some jostling between them, which jostling tends to coalesce many of them into much larger drops. These larger drops, if the updraft is not too strong, will begin to fall through the cloud as rain. If the upsurging cloud droplets reach a temperature as low as +5° F. (-15° C.), which they soon do in a

by high dewpoint, say more than 60° or 70°; and steadily decreasing surface pressure during the late morning and early after. noon. Another useful early-afternoon indication is increasing 'crash' static from a radio set — actually radio waves from lightning flashes beyond the horizon. Finally there are the definite evidences of frontal activity and advance towards one's position to be obtained from some sort of weather map — even a simple regional map that anyone can draw from airways radio weather reports in a few minutes' time.

One or more of the indications just listed can of course be present without resulting in thunderstorms, and thunderstorms can occur when one or more of them are absent; but in general the presence of most of them makes thunderstorms likely, and the agreement of all of them makes thunderstorms certain.

Given these favorable conditions, how does an air mass thunderstorm actually form and grow? The beginning of a thunderstorm is similar to the formation of an ordinary cumulus cloud or an ordinary shower. But in the case of the thunderstorm the air is more moist and more unstable, and this moist instability extends to greater heights. Hence the updrafts are more violent, and the cloudbuilding is more rapid and extensive.

As in ordinary cumulus-building, the rising currents of a thunderstorm cause the warm, moist air to cool by expansion to its dewpoint, forming minute liquid droplets (on commonly present condensation nuclei), perhaps 1/2500 inch in diameter, which droplets appear as the white or gray opacity of the cloud; and this condensation of water vapour, in a vicious cycle, releases latent heat which helps to drive and intensify the rising currents, which condense further moisture, and so on. The result is that under favorable conditions the thunderstorm grows fast and far – perhaps up to altitudes of six or seven miles.

Just as in the formation of all rain, so, while the minute cloud droplets of a thundercloud are carried up in the rising currents, there is some jostling between them, which jostling tends to coalesce many of them into much larger drops. These larger drops, if the updraft is not too strong, will begin to fall through the cloud as rain. If the upsurging cloud droplets reach a temperature as low as +5° F. (-15° C.), which they soon do in a towering cloud (at a height of four or five miles in temperatezone summer), some of them are jostled into ice particles; and ice particles are perhaps also subliming directly out of water vapour (onto sublimation nuclei already present at these cold heights), if the temperature and humidity conditions are favorable. In a fully developed thunderstorm the ice-particle cloud is often visible as a cirrus-like, anvil-shaped top or cap that blows out somewhat ahead of the main cumulus turret on the upper winds. The myriads of ice particles, falling slowly and growing into larger crystals, eventually produce rain and other forms of precipitation.

In this fashion the production of rain continues until large drops perhaps a quarter of an inch in diameter are falling through the cloud; and the rising currents build up until, in the forward and most active part of the storm, they exceed twenty miles an hour. This updraft will just support a three-eighths-inch raindrop; and no raindrop larger than three-eighths inch can exist in nature, for it would be instantly blown into smaller drops by the wind of its fall. Thus no rain can fall through the most active updrafts of the thundercloud, and the active front of the storm is marked by an abrupt stoppage of large-drop rain falling from above, by the deflection of this rain upward towards the rear of the storm, and by the breaking up of large drops into smaller drops.

From the breaking up of these water drops, which can, of course, recombine above and descend to be broken again, come the electric charges that produce the lightning. At the instant of spattering, the to-be-broken drops are charged with positive electricity, and the air surrounding them is negatively charged. This negatively charged air is carried away into upper and rearward portions of the thundercloud, but the positively charged raindrops, breaking and recombining, tend to accumulate in the most active (forward and lower) portion of the storm, which hence becomes a region of high positive electrical charge.

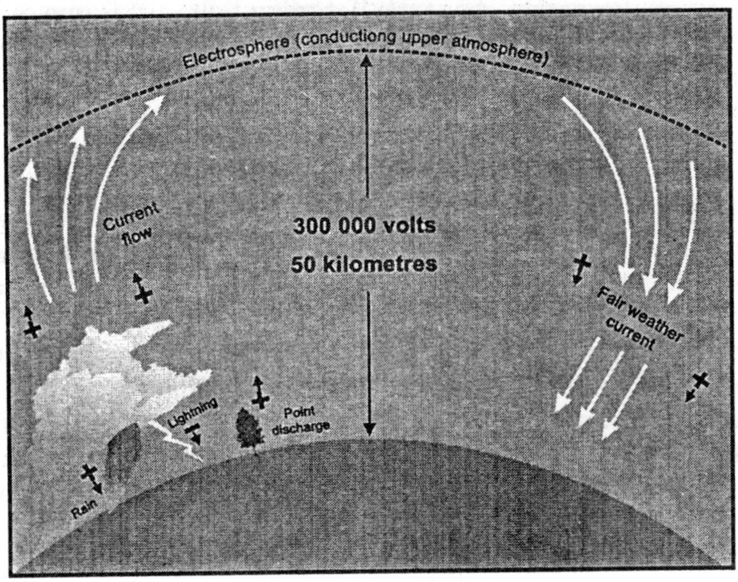

Fig. Electric Fields and Lighting Flashes

From this positively charged region, usually just above and behind the roll cloud or 'squall head,' frequent heavy thunderbolts strike downward and forward directly into the ground (whose charge is neutral or somewhat negative), while other longer but less intense flashes strike upward and backward to lose themselves in the upper immensities of negatively charged cloud. Towards the rear of the storm, however, where the earth is positive with respect to the negative clouds above, less intense bolts strike upward from the ground into the clouds.

Lightning may be only one feature of a great storm perhaps marked also by cloudbursts, ice showers, and destructive blasts of wind, but it is spectacular enough, and under some conditions hazardous enough, to warrant examination in some detail before considering the other aspects of the storm.

THOR'S HAMMER

The potentials set up by the cloud-formed electric charges of a thunderstorm are enormous — in the neighborhood of one hundred million volts — and they discharge through dry air, ordinarily considered one of the best insulators, to usual distances of half a mile, or even occasionally to a mile or more. The lightning bolt itself, though similar in many ways to a small spark in the laboratory, yet shows additional and peculiar characteristics introduced by its enormous voltage and the great length and diversity of its path. There is nothing oscillatory or alternating about the lightning bolt — it is one or more surges of current in the same direction. Moreover the bolt does not jump instantly from a positive cloud, say, to the negative earth — it appears to grow or burrow through the air in a succession of electronic darts, building its own ionized (or electrically conducting) path as it goes.

Ordinarily this growth is frightfully rapid, and takes place in perhaps one millionth of a second. But occasionally it is slow enough to be seen by the naked eye in the form of 'rocket lightning,' in which a growing and perhaps upward-directed bolt burrows visibly at no greater speed than an ordinary rocket. Some of the veined lightning in a detached thunderhead, viewed from several miles away on a clear night, can often be seen to grow or burrow in this fashion progressively along the bulges of cloud. Usually lightning strikes from thundercloud to ground, from ground to cloud — or, even more commonly, between clouds. But on rare occasions it flickers or glows out of a clear sky, as charged masses of air pass near each other or the ground.

The amperage or rate of electric current flow in a thunderbolt is likewise enormous, compared with most man-made currents. Its value ranges all the way from ten thousand amperes in weaker flashes up to perhaps two hundred thousand amperes in larger bolts. One hears of people being struck by lightning and surviving the ordeal, but these fortunates are probably grazed by side flashes or forked branches while the main bolt goes elsewhere. Few if any animals, human or otherwise, have ever been hit by a heavy main bolt and lived to tell the tale. For like its voltage and amperage, lightning power is unbelievably large — perhaps ten million million kilowatts or thirteen trillion horsepower! But as an average single flash lasts only three or four millionths of a second, the total amount of electricity (or electrical energy) involved is not very large — perhaps three thousand kilowatthours or so, obtainable commercially within more usable limits of current and voltage for seventy to a hundred dollars.

Lightning punishment at one point is by no means limited to a single bolt, however, and the old adage that 'lightning never strikes twice in the same place' is about as untrue a statement as anyone could conveniently concoct. Many lightning flashes are of the 'multiple' variety; the bolt not only strikes twice, but may strike repeatedly — five times, ten times, or even forty times in the course of a second or two. This multiple lightning usually appears to the eye as a single, flickering, long-lived bolt, and it is the kind most likely to start fires in any dry and combustible material along its path.

The brilliant flash that accompanies a lightning discharge, formerly thought to be caused by sudden heating of air, is now known to result from ionization of the air by the terrifically strong electric field along the path, causing the very atoms of dense air to vibrate at such high frequencies that they radiate visible light in the same way that rarified neon in a glass tube glows under moderate electrification. The lightning flash is thus kindred (though once or twice removed) to the flickering aurora of polar skies, and to the glowing message of an advertising sign. Lightning is usually white in colour, combining the spectrum of nitrogen with that of oxygen. But it may appear bluish in contrast to yellowish artificial lights, and may appear yellowish if viewed through haze, fog and clouds. Occasionally a reddish or pinkish flash occurs, when air rich in water vapour is ionized so as to give the spectrum of hydrogen.

Thunder takes many audible forms, but they all originate in one underlying cause — the sudden increase of air pressure along the path of the bolt (for any sound wave is only a sudden increase in air pressure) resulting from the heating, molecular dissociation, and ionization of air by the bolt. Earlier vague assertions about 'air collapsing to fill up the hole hollowed by the bolt,' and such notions, are now in the discard. If lightning strikes close enough — within a few feet of us — the sound is simply one sharp and terrific 'bang!' like that of a large cannon fired in the same place; for the ear is too paralyzed by this first and nearest blast to notice the subtler and later gradations of reverberating thunder arriving from more distant portions of the bolt.

But if the bolt is farther away, and particularly if it is a long one, the thunder arriving successively (at eleven hundred feet per second) from its more and more distant portions has the sound of a long roll, which wavers considerably even as the characteristics of the bolt, and its direction with respect to the hearer, vary greatly along its path. Under certain conditions of air stratification, and often in the presence of steep mountains and cliffs that also serve to reflect and intensify sound, the roll of the thunder echoes back and forth; these overlapping echoes, together with simultaneous sound reinforcements from equidistant portions of the path, may build up into great booming crescendos of noise.

Thunder is seldom heard over horizontal distances of more than fifteen miles; and if the flash occurs at altitudes of more than two miles, it may not reach the ground at all in audible form.

Several kinds of lightning have been observed, of which by far the most common is 'streak' lightning — the ordinary bolt. A streak of lightning is not a zigzag, as commonly drawn, but sinuous in shape, like a river in broken country. The streak may be single, but more commonly, part of it splits off into smaller downward branches that may or may not reach the ground. Occasionally a bolt splits into 'forked' lightning — two or more main branches that each strike home. Multiple streak lightning, drifted sideways by a strong wind, sometimes appears as 'ribbon' lightning. And if a single streak follows a very sinuous and varied path, it may present the illusion of glowing beads where it is either more intense, or else where part of the flash is seen end-on.

Another distinct variety, very rare but very dangerous, is 'ball' lightning. Scientists long doubted altogether the existence of this variety, but so many accounts of its appearance were authenticated that it has come to be believed in, and more or less explained. It appears in the form of a luminous ball, perhaps the size of a clenched fist, which either falls through the air, floats horizontally, or rolls along some solid object such as the ground. Visible only for a second or two, it disappears either quietly, or with a terrific explosion. According to present theory, ball lightning is nothing more than a creeping corona discharge along a relatively intense portion of the electric field — or, one might say, a 'stalled thunderbolt,' similar in kind to rocket lightning but much slower. If the electric charge behind it dissipates, the ball fades away also; but if the original charge increases enough to break down the whole path with a large thunderbolt, as it well may — then look out!

Another variety of Nature's electric illumination is 'sheet' lightning. This is not to be confused with so-called "heat' lightning — the diffuse and silent reflection of distant electric flashes hidden either by clouds or by the horizon. True sheet lightning occurs as a momentary glow extending through fairly large masses of air in the cloud region (or troposphere), and is nearly always white in colour. It is probably an ionization effect similar in principle to the aurora, though the sheet-lightning glow occurs on a much smaller scale than the aurora, and at much lower levels in the atmospheric ocean.

Another and somewhat similar variety of thunderstorm light is 'Saint Elmo's fire,' a bluish, brush-like discharge that sometimes appears on rooftrees, tall steeples and chimneys, the mastheads of a ship, or any other high, sharp-pointed object during a thunderstorm. It is merely the well-known corona discharge, reproducible at will in the laboratory or on a high-voltage transmission line.

Occurring in nature, it indicates intensification of the electric field, and may indicate the imminence of a thunderbolt. Such a corona discharge, of course, presents a very real fire hazard — particularly in the case of a hydrogen-filled balloon or airship. On May 6, 1937, during a great line squall over New Jersey, the German airship *Hindenburg* was valving explosive hydrogen in preparation for a landing at Lakehurst. Saint Elmo's fire ignited the escaping hydrogen-air mixture, and the great airship crashed in flames, killing thirty-six of its transatlantic passengers. This incident marked the tragic end of hydrogen-filled airship activity; and as the United States Government wisely declined to release helium to a nation arming to the teeth for a treacherous and unprovoked attack on its neighbors, it marked the effective end of all German airship activity for some time.

Chapter 8

Lightning Hazards

Just as lightning bolts themselves exhibit almost endless varieties of form and intensity, so do their earthly effects vary greatly, not only with the bolts' severity, but even more with the things and substances which a ground-strike encounters on its path to earth. Metallic conductors of sufficient size — say a bar of copper or other metal the size of a pencil — pass the bolt without harm to themselves or anything else. But if the con- ductor does not offer the shortest and easiest path to earth (from the viewpoint of the hop-along bolt itself), it may jump to a more tempting and destructive path; and even if the main bolt follows the conductor, side flashes may jump to neighboring conductors, be they metal tanks or pipes, pools of water, or human beings.

Small, hollow conductors may be crushed by the intense electrostatic and electromagnetic fields accompanying lightning passage. Irregularly shaped, loose, or insecure rods or wires are likely to be moved, bent, burned, or melted; and insufficiently large conductors, such as a radio antenna, are invariably melted or vaporized instantly by a direct stroke.

Non-conductors such as wood and stone fare much worse, at least during the 'dry' part of the storm before they are wetted over with a film of partly conducting water. Dry trees and poles are often split and splintered throughout their length by the explosive action of their moisture content suddenly converted into steam, and the pieces are blown forcibly outward in all directions. Dry or partly moist bricks and mortar may be violently exploded, and very rarely a small and flimsily built house is wrecked completely, as if by an earthquake.

Some of lightning's worst destructive effects are seen where large conducting masses are separated by non-conducting ones -such as a metal roof, ungrounded, on top of a wooden house. Here the conductor seems to collect and intensify the bolt, which wreaks havoc on the non-conducting structure beneath. No ordinary insulator such as a glass plate, of course, offers any effective resistance to a lightning bolt. If an insulating plate stands in the way and cannot easily be got around, the bolt simply blasts or burns a hole right through it.

The best possible form of lightning protection for any object, animate or inanimate, is a 'Faraday cage' — in complete form a grounded shell of metal not less than one twentieth inch thick (Number 16 A.W.G.), completely surrounding the object in question and several feet away from it on all sides. But as such a metal box would be very expensive and inconvenient, it can be approximated, with nearly as good practical results, by a skeletonized cage of interconnected conductors — each conductor being at least one fourth inch thick, if of copper; or one third inch thick, if of iron.

Of such general form are the lightning-rod systems now in use on thousands of isolated houses and barns in thunderstorm areas — systems that prove their usefulness by greatly reducing fires and other lightning hazards. In general, these protective rods should extend at least a foot or two above all the higher projecting parts of a building; they should extend downward as directly as possible to thoroughly good ground connections; and they should be bonded to each other and to any other large metallic masses in the building, which latter should be independently grounded.

Any high metallic conductor, such as a steel tower or a steel-frame city skyscraper, usually confers about it a 'cone of protection' having a radius equal to three or four times the height of the tower; occasionally, however, this cone of protection is violated by an erratic bolt. But for all the unpleasantness of lightning in the open — the vivid flashes of uncomfortably close ground-strikes and the deafening crash of near-by thunder — the actual hazard is really very small. Out of all the millions of people exposed to lightning in the United States, only about five hundred are killed by it each year. Under the intense part of a very severe storm, however, and in various unwise locations beneath any storm, the risk may notably increase. The Bareau of Standards has issued some rules of wise personal conduct for those caught out in the open by storms, which are as follows:

- Do not go out-of-doors or remain out during thunderstorms unless it is necessary. Stay inside of a building where it is dry, preferably away from fireplaces, stoves, and other metal objects.
- If there is any choice of shelter, choose in the following order:
  - Large metal or metal-frame buildings.
  - Dwellings or other buildings which are protected against lightning.
  - Large unprotected buildings.

  - Small unprotected buildings.
  - If remaining out-of-doors is unavoidable, keep away from:
  - Small sheds and shelters if in an exposed location.
  - Isolated trees.
  - Wire fences.
  - Hilltops and wide open spaces.
- Seek shelter in a cave, a depression in the ground, a deep valley or canyon in dense woods, or a grove of trees.

Their cumulus clouds are merely prosaic banks of condensers, but their lightning is real and convincing, as anyone can testify who has seen one of their fivemillion-volt flashes strike downward in the dark through fifteen feet of air to shatter a three-inch fence post with the bang of a small cannon.

Safety of People in Ordinary Wood-Frame Houses

Much will depend upon the construction of the house and what is in it, and particularly the amount of metal in the walls, whether in the form of gas, water, or vent pipes, how the electric wiring is arranged — whether in rigid conduits or in BX, or knob-and-tube open wiring. Quite obviously, the only absolutely safe place to be is inside of a completely enclosed metal structure, which may or may not be grounded. The degree of safety enjoyed by the occupants of an ordinary wood-frame house will depend upon the degree of approach to this completely shielded structure. The wave front of current discharge in a direct stroke is steep enough, and the magnitude great enough, so that in many cases fairly high potentials will be built up in conductors of thirty- to forty-foot lengths — so that side-flashes may occur to other metal structures which may be connected to the same ground but at some distance away.

This was particularly exemplified in a case where a man had a lightning rod system properly installed, as far as the lightning rods were concerned, and yet it flashed through about six inches of wood and air to a conduit located in the peak of the attic roof underneath the lightning rod system.

Of course if these two had been tied together at this point, the fire which resulted would have been avoided. So far as my experience goes, the greatest danger from lightning in the ordinary wood-frame house is from fire, and of course this is particularly so in the rural communities where fire-fighting apparatus is not readily available. In cities the damage done to buildings as a result of lightning is not as a rule serious, although cases are on record where explosions have occurred during lightning storms, but apparently these explosions were due to the existence of gas or other explosive materials enclosed within the house at the time it was struck.

Wood-Frame Houses Protected by Lightning Rods

Insurance records for the State of Iowa indicate a very substantial improvement, which is also substantiated by those from the Province of Ontario, as reported by the Fire Marshal of Ontario. These results have indicated that lightning protection is better than ninety per cent efficient, from the point of view of preventing fires in protected buildings.

Of course not all of the lightning protection systems were properly installed. Many were, no doubt, inadequately grounded, which accounts for the imperfect record to some degree. At the same time, we know today that lightning discharges do not always strike the highest object, and that now and then the cone of protection, which laboratory tests have demonstrated exist about a lightning rod, is violated. I have, for instance, a photograph showing lightning striking the top of the Empire State Building, in which the discharge failed to contact the very top of the building but struck about forty feet down. It is apparent that lightning discharges take a path which might well be called the best path, from the point of view of the discharge itself.

Safety of People in Closed car with steel Turret top

Following the principle of the Faraday cage, there is no reason why people inside of a car with a steel top should not be very safe from lightning discharges, provided that they are not in the process of getting in or getting out of the car, or do not have their heads or arms out of windows, or are not standing on the ground changing a tire, and assuming that the gas tank is not struck and punctured, which is relatively unlikely. I believe the best advice to a man riding in an automobile during a lightning storm is to keep going and not to get out of the car, and the chances are that he will be perfectly safe. Of course, if power wires fall on the car, which has happened during violent storms, he is wise to remain in the car, and not attempt to get out, since this procedure has in the past resulted in serious accidents.

Safety of People in Closed Wooden Yacht

As a general thing, lightning is more likely to strike a wet mast on the water, if it is striking round about, than a similar object on land, because of the relative smoothness of the water and the relative height of the projecting mast. It is probable that suitable protection could be provided to the boat and its occupants by connecting a conductor following down the mast to the metal keel of the boat. In case it is an engine-driven boat, care should be taken to be sure that all the metal objects within the boat which are not connected together are electrically bonded together, and connected to some outside metallic conductor, such as the keel, which furnishes good contact with the water.

I don't know of any definite rule which has been set up concerning the protection of boats, but it seems to me that the general principle of the Faraday cage, or its equivalent, is one which can be applied to boats, coupled with the idea of the lightning rod and its usual cone of protection, and means for discharging the current of the lightning discharge itself into the water, without having either the boat or its occupants become a part of the circuit. It might be well at this point to inject an observation concerning people in swimming, to the effect that, in my estimation, the danger to the swimmer is not so much that he may be struck by lightning, as that a small current flowing through the water, even at considerable distance from the stroke itself, may be sufficient to render him unconscious, and he dies by drowning, rather than by electrical shock.

Safety of People in an Airplane

Once more the question is one of preventing the passenger or pilot from becoming a carrier of current of the lightning discharge by being in series with it. The only known way to prevent this is either to keep the lightning current from entering the plane by means of an all-metal plane, or its equivalent protection by some means of shunting the discharge which gets into the plane from the body of the occupants, and means also of preventing the lightning discharge from interfering with the mechanical and electrical operation of the plane. The best method of doing this is probably the all-metal plane, which brings us to the Faraday cage, where if the joints in the plane are thoroughly bonded together, and the metal parts so connected that the measuring instruments and other electrical circuits of the plane are not likely to become a part of the lightning path, the possibility of damage is relatively small. The occupants of the plane ought to be perfectly safe provided they are entirely surrounded by metal having a reasonable current carrying capacity....

From the point of view of fire, the multiple stroke is undoubtedly far more serious than the single stroke, and I believe I am quite safe in stating, that the single stroke of large magnitude is probably responsible for the explosive effects frequently observed, whereas the multiple stroke, even though of many relatively small currents, may be responsible not only for explosive effects but for fires also, provided inflammable material is present. As many as forty strokes have been observed within one second of time, and data which we have been taking during the past two years seem to indicate by far the great majority of lightning strokes are multiple....

DOWNBLAST AND DELUGE

Earlier in this chapter we saw how a typical thunderstorm forms out of nothing into a towering Everest of threatening cloud, how the terrific electric potentials of the lightning arise out of a mere spattering of raindrops. But what about the lashing hail, the scouring squall winds, and the other thunderstorm accompaniments that often prove more dangerous and destructive than the lightning itself?

We had earlier in mind the picture of a towering cumulus cloud marked by fast-rising air currents and fast-falling raindrops of near-maximum size, whose spattering in the zone of strongest updrafts produced positive electric charges large enough to forge the thunderbolts. As most of the warm, moist air contributing its latent heat to these rising currents is coming from in front of the storm, the most violent updrafts are usually in the front part of the main thundercloud; and they tend to slope somewhat towards the rear of the storm, carrying the rain rearward also. Just behind the main updrafts, where their speed has fallen off to less than twenty miles per hour, will be the region of heaviest rainfall.

Much of this heavy rain, coming from upper and colder re- gions of the thundercloud, is cold; some of it may even be frozen. This cold deluge cools the air as it falls, partly by its own coldness, but even more by its partial evaporation. The result is that air in the heavy rainfall region is soon chilled well below the surrounding air, hence becomes denser than the surrounding air, hence begins to sink, and soon develops into a great downdraft of cold air perhaps nearly as violent as the updraft of warm air in front of it. This cold downdraft reaches the ground with the forward speed of the whole storm at upper levels (which may be twenty or thirty miles per hour or more), and at the ground level it sweeps forward as a violent outrushing squall wind, blowing in heavy gusts directly out of the storm.

But the squall wind may far exceed the speed of the storm itself. It is not impossible for a storm moving at thirty miles an hour, say, to produce a surface squall which (however shortlived) may blow at sixty or eighty miles per hour and sweep all before it. These excessive squall winds are perhaps partly explained by the crowding of the downdraft against the ground (which also contributes to the sharp barometer rise noted with the squall).

But the chief reason for their ferocity is the action of a horizontal vortex. With a great updraft in one part of the cloud and a great downdraft behind it, all sorts of eddies and minor whirlwinds are bound to exist in thunderclouds, as airmen who have endured the blind bumpiness inside one can fervently testify. Most of these vortices, being well up in the cloud, are invisible (except on the gyrating instrument dials of an unhappy airplane); but one of them, being formed just at the cloud base, where the relative humidity is such that the slightest drop in pressure condenses moisture, may plainly appear as the long, light-colored roll cloud or 'squall head' arching horizontally across the front of the storm.

As the roll-cloud vortex revolves, it moves across land or water like the revolving brush of a gigantic carpet-sweeper, piling its speed of rotation on top of its speed of horizontal motion – all to the consternation of squall-blown humans in one case, or insects swept out of a carpet in the other. This horizontal vortex of the roll cloud is, in fact, a sort of horizontal tornado; and under certain conditions, parts of it can turn down to the ground and become real tornadoes. In any case, it is something for an aviator to avoid at all costs.

This whole thunderstorm wind-mechanism picture, has, of course, been snapped at a single and very typical instant. Actually, thunderstorms and line squalls vary greatly in size, form, and intensity. Some thunderstorms have no squalls; some squalls have no thunder; and any one storm is constantly changing, from moment to moment, its form and intensity. Updrafts and downdrafts alike are constantly changing their speed and their position in the storm.

Thundershowers are usually of the brief, heavy type, and may reach cloudburst intensity. Raindrops in the active parts of the storm are much larger than ordinary raindrops, being mostly of near-maximum size; and they are crowded together, in certain places, by the interplay of violent and ever-varying air currents. It is not at all uncommon for these thunderstorm deluges to fall at the rate of three inches or more an hour, flooding low places with a few feet of suddenly rising water.

On occasion, the rate of water-fall for a brief interval has been known to run as high as one inch per minute! To a ground observer such a cloudburst seems to be falling in sheets of solid water, as if the heavens had indeed opened; to an aviator trying to fly through the downpour, it seems almost like boring through the depths of the ocean, and airplanes have actually been brought down by running unsuspecting into a heavy cloudburst at night. On the ground a cloudburst is most destructive in normally dry desert and mountain regions — among steep converging slopes bare of brush or timber — where any dry ravine may within a few minutes become a raging torrent, and walls of tumultuous water ten or twenty feet high may roar along the rock-walled bottoms of steep canyons.

Nor is torrential rain the last of possible thunderstorm visitations on men caught in or below the storm. On occasion, thousands of tons of ice may fall from the skies in the destructive form of hail. As we have seen earlier in this chapter, the upper parts of a towering thundercloud harbor many slowly falling ice particles. Under favorable conditions of temperature and humidity, some of these may grow into sphero-crystals or snow balls (*graupel*) about the size of small peas. Alternatively, or in addition, raindrops carried by the storm's great updrafts into temperatures well below freezing, and sufficiently jostled by turbulence, may be frozen into ice balls (sleet) of similar size. When the updraft becomes less violent, or when these pellets reach the edge of it, they fall groundward again, perhaps gathering a coating of snow above the freezing level, and a further coating of slush or water below it.

Soon, however, the growing ice pellet may enter new violent updrafts, which carry it upward again into the arctic zone, where its acquisitions of snow, slush, and water are frozen into a new concentric shell of solid ice, increasing its diameter. In like manner the ice pellet may fall again through several thousand feet, be carried upward again through another great rise, and repeat the whole down-up cycle ten or twenty times, growing the while into an ice ball of considerable size.

The largest hailstones ever authentically observed, which fell on some Nebraska farms in 1928, were the size of small grapefruit. One of these stones weighed one and one half pounds and was about six inches in diameter! Three-inch hailstones, the size of ordinary oranges, probably shower down somewhere in the United States a few times every summer. Stones of golf-ball size are not at all uncommon. In 1936, nineteen Bushmen in South Africa were reported to have been killed by a fall of oversized hail that covered their bodies to a depth of three feet. Hail of some sort accompanies a large proportion of temperate.

one thunderstorms, but the hail zone of any one storm is usually not more than a few miles wide. But in tropical thunderstorms hail is very rare, however heavy the lightning and rain, because rain-making portions of the thunderclouds seldom extend far into a freezing zone perhaps five miles high.

The higher a temperate-zone thunderstorm extends, the blacker are its underparts, and the more likely it is to unleash severe lightning, torrential rain, destructive winds, and hail. The convergence of two or more storms tends to intensify these factors. Wind-tunnel tests with hail-sized balls have revealed some astounding probable speeds for the updrafts of great storms.

To sustain a one-inch hailstone an updraft of about sixty miles per hour is necessary; a three-inch stone requires a hundred-mile current; and a five-inch stone must have been held aloft at some time by an upblast of two hundred miles an hour! In severe thunderstorms, therefore, 60-mile updrafts (and commensurate adjacent downdrafts) are common, while 100-mile and 200-mile currents are not impossible. Which seems good enough reason for aviators to keep out of them if they can.

An average detached or 'air-mass' thunderstorm starts as a pile of cumulus clouds perhaps five miles or so in diameter, gradually spreading out horizontally and building up vertically as it sucks in warm surface air and gathers to itself near-by clouds. The altitude of its highest cumuliform turrets may be four or five miles in temperate latitudes (or considerably higher in the tropics); and the cirriform anvil or 'cap' may reach six or seven miles aloft. As the storm moves bodily along with the upper winds, it spreads out laterally to a width of perhaps twenty or thirty miles, so that its whole path resembles a pear in shape. After six or seven hours, with the decline and disappearance of the sun, an air-mass storm usually dies out, having raged in all through a distance of perhaps a hundred and fifty or two hundred miles.

The airplane has of course made it possible to survey the breath-taking topography of thunderstorm clouds, quickly and accurately. One of the TWA 'over-weather' research aviators made a detailed investigation of a large air-mass storm above Missouri. The general storm-top of cumuliform clouds was at twenty-two thousand feet. The highest cumulo-nimbus turrets boiled up to twenty-five thousand feet, at which altitude the plane could just clear the turbulent wind-shift line. The cirriform anvil, ten miles wide, extended on up to about thirty-five thousand feet. Flying through the anvil at twenty-six thousand feet, the plane encountered fine snow and some subcooled water droplets which froze as light icing; the temperature was 5° F., compared with 100° F. at the ground.

Even an ordinary thunderstorm has within it a cold front in miniature — a rain-cooled wedge of squall-wind air surging outward and displacing the warm air in front of the storm. But the cold air behind a large-scale front usually comes from arctic or polar regions, whereas the cold air in a thunderstorm is, one might say, manufactured locally by the refrigerating process of the storm. Yet if a very large thunderstorm spreads out laterally beyond fifty miles or so, it begins to take on the characteristics of a line squall.

THE WINGS OF THE STORM

Reduced to typical form and simplest terms, a line squall is nothing more than a violent cold front often spotted with many thunderstorms, continuous or intermittent, along the hundreds of miles of its length. The front itself is necessarily marked by more than ordinarily large temperature contrast between the unlike air masses; by high moisture content and great depth of the warm air mass; and perhaps also by some moisture and instability in the cold air mass. Also in evidence is the rapid advance of the cold air mass, the enforced retreat (mostly sidewise) of the warm air mass, and conflicting impact between the two. This air-mass conflict shows up as high winds along and across the front, sharply contrasted in direction, and appears on the weather map as isobars packed close together and bending in a sharp V where they cross the front.

From this typical picture as a norm, individual line squalls of course vary widely. Few cold fronts are violent enough to be spectacular line squalls, and few of these, even, conform strictly to type. The one common characteristic of all surface cold fronts, at least over flat country or the ocean, is a marked shift in wind direction, usually from some southerly to some northerly quarter. The squall line itself usually advances from the northwest or west (less frequently from the southwest) at something like twenty to thirty miles per hour in summer and thirty to forty miles per hour in winter. (Occasionally, in some parts of the country, the squall may come out of the north or even the northeast.)

But rather than further generalities, often violated, here are accounts of two actual line squalls that struck the city of Washington. One came early in June:

At 2 P.M. the sky was still clear, but hazy; and a few clouds were rising in the distant west. The wind was moderate out of the southwest, and some twenty or thirty small sailboats were racing on the wide Potomac River, which was also dotted with various other craft. Upper clouds gradually grew out of the west, gradually spread and darkened, obscuring the sun in a quarter of an hour. Then a storm collar or squall cloud appeared at lower levels out of the southwest. It consisted of dark gray stratocumulus clouds, whose violent agitation showed the ferocity of their inner winds.

The squall cloud came rapidly on until it passed overhead, but there was no lightning and no rain. The wind slackened somewhat directly under the squall cloud, becoming rather puffy, and the sailboats continued their gay progress all unheeding. But perhaps five minutes after the squall cloud had passed overhead, the surface squall struck very suddenly and fiercely out of the southwest. As the wind rose shrieking to fifty or sixty miles per hour the sailboats, one after another, were pressed down to the water and capsized, some with sails still set, some under bare poles. The extreme wind lasted only three or four minutes, then moderated to twenty miles an hour for another half-hour, with light rain; after which it fell nearly calm while the rain continued.

Another squall came in May, when the late spring contrasts between tropical and polar air were most intense: Up to mid-afternoon the sky was perfectly clear but very hazy, the wind moderate southwest. About 3 P.M. the forerunner clouds appeared out of the northwest — perhaps a sort of upper-level squall head — arching across the sky from northeast to southwest. They appeared as a solid wedge of high strato-cumulus or low altocumulus, advancing steadily at an altitude between one and two miles. There were moderate puffs of wind out of the northwest, followed by a gusty northwest wind of twenty-five or thirty miles per hour — but for the moment, no rain. The high clouds swept by overhead and out to eastward, shutting out the sky, and for twenty minutes nothing more happened, except that the pressure jumped up by five hundredths as the cloud edge passed, and the gloom steadily deepened in the northwest.

Then the gray rain curtain appeared, and with it came the real squall at 3.50 P.M. For a few minutes the northwest wind whistled around sixty-five miles an hour, driving torrential rain and rattling hail horizontally against windows and doors. Lightning flickered incessantly above and banged heavily to ground, trees and poles were borne to earth in exposed places, and the horizontal rain flooded over everything, cutting visibility to zero. The base of the black thunderclouds, when they could be seen at all, looked about one thousand or two thousand feet high. The temperature fell twenty-four degrees in an hour. The worst of the squall was over in a few minutes, but scattered showers and mild thunderstorms continued, interspersed with broken clouds, throughout most of the evening in the unstable invading cold air mass.

These are the picturesque and transient line squalls of spring and summer. In winter, marked cold fronts are less spectacular, but if anything more unpleasant. Their approach is more rapid, and the cold air mass is much colder. Intense winter fronts are likely to be marked by a deep zone of blizzard-like snow.

The worst weight and hazard of summer line squalls comes on the water, as the sailboat episode above indicates.

FLYING THROUGH A LINE SQUALL

Full-bodied line squalls are colorful enough, viewed from the ground or the water. But the way to see and feel one at its best — or its worst — is to get up into the air where we are almost a part of the gyrating squall winds. This is something that wise aviators seldom do, intentionally (except in big, scheduled transports, or in urgent military flying); but they sometimes get caught in squalls or plow through them deliberately in the course of an urgent mission. On the same afternoon that the airship *Hindenburg* crashed in flames at Lakehurst, an army pilot and I were flying from Boston to Washington in *BT*-8 — a medium-fast, stubby, powerful pursuit-type training monoplane:

... Towards Trenton the clouds were more scattered, and we dropped down through them to 2000, heading out westward so as to pass well to windward of Philadelphia's smoke pall. The scattered cumulus clouds, now above us, were becoming larger, with some scattered showers among them. Some of these shining white cumulus heads were beginning to mushroom quite rapidly up into the blue — a sign of thundery weather.

Soon there appeared far ahead, somewhere over northeastern Maryland or southeastern Pennsylvania, a dark line or roll of cloud arching across the entire sky from southwest to northeast. This was the real front, the line squall itself. Great white thunderheads towered along it, in some places much higher than others. Beneath these higher cloud-peaks were gray curtains of rain, and around them flickered the seeming-miniature sparks of lightning. For the time being, the controls were in my not very experienced hands.

Our course was about 240° magnetic (southwest-by-west true), and as we approached to within ten miles or so of the squall line, made out two large thunderheads towering far above the long, continuous roll cloud — turreted white mountains three or four miles high. Under each thunderhead a dark deluge of rain was falling from the squall cloud to the ground. It seemed wise to keep out of these active storms if possible, and we changed course 15° to the right so as to fly midway between them, just under the long, black roll cloud (which appeared fortunately to be well up off the ground, at a height of half a mile or more).

Beneath this rapidly approaching black roof of turbulent cloud the visibility was good — we could see far out ahead into patches of sunlight and shadow beneath other clouds behind the front. We were still cruising in level flight at 2000, and the black ceiling at about 3000, now rushing towards us overhead, seemed to offer plenty of clearance.

As we passed beneath the dark cloud-roof the *BT*-8 began to bounce around crazily. I had expected bumps, but these were so sharp and ferocious that it was all do to maintain oncourse control of a plane that was, at best, rather tricky to fly. Snatching a glance at the instruments, was alarmed to see the rate-of-climb meter swing to plus 1000 ft/min (12. m.p.h. straight upward), in spite of the fact that was already nosing downward in an effort to get lower and escape from the wildly gyrating black festoons of cloud that seemed now just over our heads.

This black cloud-roof, marked by turbulent downward bulges of mamato-cumulus and momentary 'cloud spouts' or incipient whirlwinds, was getting nearer by the second. It seemed as though we were about to get sucked bodily up into its black chaos. I had the nose shoved down as steeply as seemed prudent, for the air speed was now over 200 m.p.h. and the terrific bumps might strain the wings at this speed. Yet I hesitated to change the throttle setting without instructions from the pilot.

Finally I jammed the nose down in a steeper dive, taking a terrific beating from the bumps and still not losing altitude. At this point the great down-bulging cloud-pocks seemed to be almost couching our heads. Just then the pilot took over, and throttled back to case the bump-strains. And within another minute, we had passed out from under the roll cloud into a region of scattered sunlight and shadow, marked by detached storms in the cold air mass. These detached storms, of course, could be easily flown around....

Passing through a line squall in a controllable airplane may be exciting enough, but it is relatively easy and pleasant compared to navigating through one in an uncontrollable free balloon. Lieutenant Settle of the United States Navy had this unenviable experience over western Pennsylvania on the twentyeighth of May, 1938. Wrote he, in part:

Dropped ballast and at 18.15 (6.15 P.M.) balloon was at 4500 feet altitude; the line cloud overtook and passed under us and we started up rapidly, the variometer going 'hard over' on the ascent side. At 8300 feet I valved two seconds, and balloon leveled off at 8400 feet and then started down at a violent rate, the variometer ink going out of sight on the descent side. While falling the air was extremely turbulent; hail (size of peas), heavy rain, and snow flurries hit us; the basket swung violently from side to side like a pendulum and spun in azimuth; heavy gusts hit the envelope and basket; the motion of the balloon was like when attempting to hold one on the ground in very gusty surface wind. Lightning and thunder were heavy and apparently right on top of us. The rain and air were very cold and our hands numbed.

During the descent we dropped approximately fifteen bags of ballast; we came through the clouds at 3000 feet and leveled off at about 2000 feet. Found myself under a dense black cloud, the line boundary of which was several miles south of me; the line extended as far as see in either direction. There were areas of heavy rain along the line on both sides of me and heavy rain to the northward of me, with lightning and thunder on all sides. The rain in my immediate vicinity had temporarily ceased; my air was NNW, about thirty miles per hour, and for a few minutes seemed relatively smooth.

It was soon apparent that we were fast overhauling the line cloud. When nearly up to the line, we again started violently up through the cloud on a second 'vertical circle,' repeating the history of the first one, and attaining a maximum altitude of 8200 feet. Some of the elements were more violent in the second than in the first excursion. Settle finally managed, by adroit maneuvering with gas valve and ballast release, to case his balloon out of the squall line. He landed at dusk on a farm near Parryopolis. He lived to write of his experience, but others who dared line squalls in free balloons, such as the brilliant American aerologist Meisinger, were less fortunate and plunged to lightning-lit deaths.

Chapter 9

Tornadoes and Waterspouts

Tornadoes are the smallest of all true storm whirls, but they are by long odds the most fearful and the most violent. The path of tornadic destruction may be only a few yards wide, and never spreads to more than a mile across — but within that path, winds roar at super-hurricane velocities approaching five hundred miles an hour. Laths are driven like spears through large trees, and straws are driven like darts through inch-thick boards. Trees are uprooted and carried away, houses are exploded and laid in ruins, débris is strewn over the countryside, and the ground itself is harried and scoured. Cattle and men are carried aloft, cut in two or spared entirely, dropped yards away or miles away, gently or roughly, dead and mangled or alive and kicking.

The basic principle of a tornado can be glimpsed in the vortices that a vigorously plied canoe paddle creates in still water -or in the circular whirl of water around a washbowl drain. For the great air whirls likewise originate as vortices between conflicting currents — currents in the flow of the upper air. These strongly opposed upper winds arise oftenest in the more turbulent portions of a large and vigorous cyclone, either temperate or tropical — a cyclone that is usually some hundreds of miles wide. Sometimes the smaller whirls are themselves incorrectly called 'cyclones,' but 'tornado' is the true name. It comes from the Spanish and Portuguese words for 'thunder,' having been first applied to twisting thunder squalls experienced by early mariners off the northwest coast of Africa.

Full-bodied atmospheric tornadoes are rare storms even in the geographic regions of their most frequent occurrence — parts of the American Midwest and Midsouth. The belt of maximum tornadic activity extends roughly northeast-southwest through Iowa and Kansas; but other midwestern and southern states also have their share. No state cast of the Rockies, in fact, is free from occasional 'twisters,' and some few have been reported along the Pacific coast. All across America the yearly twister toll is about two hundred and fifty lives, some of which could doubtless be saved by a better application of aerological knowledge.

The storm waterspout, equally rare, is simply a tornado gone to sea, with all the sucking, rending violence of a land tornado — except that it usually has nothing to tear apart except sea water, unless some mariner is unlucky enough to fall afoul of it. But the commoner fair-weather waterspout of tropic seas is a milder creature, somewhat akin to the familiar hot-weather dust whirl ashore. Basically the dust whirl is a sort of model tornado, formed in plain view in the lower air levels on a small scale. Hence, it is worth examining as an introduction to the more awesome and confused mechanism of true tornadoes.

Growth of a Dust Devil

Almost everyone has noticed the small air whirls that frequent dusty roads and dry plowed fields on a hot summer afternoon. In subtropical desert country, where summer daytime temperatures are well above a hundred degrees under a blazing sun, where dry winds ebb and flow across the flat and sandy expanses, these dust whirls grow to rather imposing size and power.

Perhaps one overheated spot, like a hot stove, starts a rising current of air. The surrounding air flows inward to equalize the pressure difference, possibly from vagrant currents that are already somewhat in conflict — blowing in opposite or converging directions, or even blowing in the same direction at different speeds. Thus is started an eddy of whirling winds that ascends in the form of a spiral. Beginning with a sudden whirl on the ground, it grows with the ballooning of its overheated air to higher levels, perhaps reaching a thousand feet or so above the desert if conditions are favorable, and moving along with the general motion of the air.

The column of whirling dust is usually not more than a few feet wide, and its progress is slow enough so that anyone on foot or on horseback can easily avoid it. But even if it hits us squarely, no serious damage is done. It will buffet us with shifting winds, fill our mouth and ears with dust, and try to snatch off our hat. Papers and light trash lying loose are carried aloft for a few hundred feet. A large tent fly may be burst upward like a paper bag. But an ordinary house stands firm. Thus the desert whirlwind is something like a tornado, at least in principle, though comparatively puny in effect.

The dust whirl is limited in size and intensity by the limited amount of energy present in its atmospheric causes. Unstable air and conflicting air currents, in some form, are responsible for all whirlwinds — dust whirls, waterspouts, and tornadoes alike. But in the case of dust whirls both these causes operate on a small and limited scale. The instability over a superheated desert may be very pronounced; the vertical temperature gradient or lapse rate may be more than about – 19° F. per 1000 feet (about -34° C/km), so that the air overturns of its own accord; and it is certainly greater than -5° F. per 1000 feet (-10° C/km), which permits overturning once begun to continue without the aid of moisture.

But this instability does not extend to very great heights, and there is practically no moisture present to contribute its latent heat of condensation to the building of large updrafts. The conflicting winds that feed into the eddy are merely local winds, close to the ground and retarded by friction with the ground. As the earth's rotation is not effective on this small scale, dust whirls may spin either clockwise or counter-clockwise, in accord with chance.

Nevertheless, a dust whirl has one quality that illustrates in miniature the underlying reason for a true tornado's destructiveness – the winds inside the dust column, near the centre of the whirl, spin with a speed far greater than the speed of the outer air currents that feed them. This effect is due to a physical law – the 'conservation of angular momentum' – which governs planets in their courses no less than winds in their wanderings. In essence, this law requires that anything circling a centre must tend to circle faster as it approaches that centre.

So much for the small whirlwinds. Now for the portents that usher in the worst among all the varied freaks of the weather.

Tornado Weather

For many years aerologists had noticed that most tornadoes appeared in the southern quadrants of temperate cyclones (more rarely, in tropical hurricanes) – not all cyclones, of course, but certain ones. The reason, now coherently explained by airmass-and-front analysis, is that certain of these southern quadrants, or 'warm sectors," provide the atmospheric conditions out of which tornadoes are most readily born. Conditions are most favorable in a 'thundery' warm sector where temperature and dewpoint are very high and where the air is conditionally unstable (having a vertical temperature gradient or 'lapse rate* greater than about -3° F. per 1000 feet [-6° C/km] with high relative humidity) to great heights. Moreover, most tornadoes seem to form somewhat (say fifty to a hundred miles) ahead of a marked and violent cold front or line squall that is invading the warm sector from the north or west.

All the conditions mentioned so far are equally favorable to violent thunderstorms, and violent thunderstorms usually (but not always) do occur along with tornadoes. The one essential condition that distinguishes a 'tornado clay' from a general thunderstorm day is strong, conflicting upper winds — blowing some forty or fifty miles an hour or more in converging or conflicting directions — at the lower cloud levels around half a mile high. High southwest upper winds are always induced in the warm sector by the crowding of the oncoming cold front, and sometimes the cold squall head, pushing eastward at say halfmile altitude above the retarding effects of surface friction, may project outward fifty or one hundred miles beyond the cold front at the ground. If so, there is formed a steep upper front between the cold and warm air. On the southeast side of this upper front the upper winds may be blowing a gale from the southwest; on the opposite side of the front they may be blowing equally hard from the north or the northeast. Under such conditions the slightest convection or updraft along the upper wind front will start a monstrous vortex between the two great slip currents.

Such updrafts are provided in gigantic form by the latent heat of condensing moisture in air that is conditionally unstable to great heights. The terrible vortex forms, perhaps a mile or so wide at the cloud level, drawing unlimited energy from the two great opposing upper air currents. Conservation of angular momentum speeds the inblowing vortical winds, not at the original forty or fifty miles an hour, but up to perhaps four or five hundred miles an hour. The vortex grows downward from the cloud level to the earth, and where it touches the earth, there is destruction.

With our present aerological knowledge, tornado weather and tornado portents still cannot be surely or accurately appraised. Quantities of radio pilot balloons and radio sounding balloons might accomplish the task. To our five senses and the old, reliable ground instruments, possible tornado weather is discerned as (1) a sticky, sultry oppressive day with southerly winds; (2) very high temperature and dewpoint; (3) pressure falling steadily and perhaps rapidly during the day. There may be, an hour or two before the tornado, a topsy-turvy sky of mammato-cumulus clouds, crazily bulging down instead of up and indicating a distinct upper layer or front of colder air. (But these clouds do not necessarily mean tornadoes.)

In the west and northwest, perhaps (if we are in for a tornado), great thunderstorms grow nearer, with lightning both severe and continuous, and some hail usually in evidence. The towering thunderclouds have an ominous appearance — their colour is often described as a sickly greenish-black — and the lower clouds are perhaps visibly in rapid and confused or conflicting motion. Then, out of the base of a dark thundercloud, the rope-like funnel cloud appears, spinning counter-clockwise in the northern hemisphere — and if this funnel hangs near-by to the southwest or west of us, it is time to go underground or to move.

The favorable times of day for tornadoes are the same as those or thunderstorms — mid-afternoon to late afternoon — although frontal or pre-frontal twisters can occur at any time. the favorable times of year are spring and early summer, when air-mass contrasts and collisions are most intense. Professional aerologists have so far had very little luck forecasting tornadoes, and they are naturally hesitant to scare everyone in a whole state when the twister, even though it occurs, will mow only one narrow swath of destruction. But it is my guess that a keen amateur aerologist in the tornado belts (or a keen professional, for that matter) could do something along this line by drawing hourly regional weather maps from airway radiophone weather broadcasts. So far, indeed, regular meteorologists have sometimes not even been able to keep track of tornadoes already formed. One that struck Hutchinson, Kansas, several years ago had been traveling steadily towards the place for five hours, yet the people of the town were not warned.

The usual height of a tornado funnel is the same as the 'ceiling' under the thundercloud base — say two thousand to three thousand feet. Since the funnel cloud results from condensation of moisture as the result of expansional cooling in the low pressure at the whirl centre, the funnel varies greatly in appearance with atmospheric conditions. Oftenest it is a sort of thin, dangling rope, sometimes a gigantic 'elephant's trunk,' occasionally a fairly wide and solid-looking funnel. Rarely (in very dry air) the funnel may be entirely invisible, though nevertheless destructive. Owing to peculiar light-refraction effects in its rarified centre, the funnel sometimes looks almost black and very solid, 'as if made of tar.' The funnel itself (or the air whirl around it) emits a roaring noise, when close by, that has been compared to the roar of hundreds of airplanes in flight or scores of freight trains going through a tunnel.

The narrow, sinuous path of destruction is usually twenty to forty miles long (though it may be as much as three hundred), and anywhere from fifty feet to a mile wide, averaging perhaps a thousand feet. The speed of advance continuously varies, being anywhere from twenty to sixty miles per hour, and the direction is usually towards the east or northeast. In addition to its forward motion, the funnel often swings pendulously from side to side, changes form and shape, and bounces up and down — all of which gyrations serve to make matters interesting for frantic humans trying to escape the dread thing. Frequently the funnel lifts off the ground for a good distance, capriciously sparing one farmhouse and blasting the next.

Tornado Havoc

The earliest American colonists were introduced, here and there, to an occasional tornado. One was reported in New Haven in 1682, and another in Charleston in 1762. In the nineteenth century settlers penetrating into the plains states were oftener exposed, but what with all the other hazards of pioneer life, twisters do not seem to have been very seriously regarded until the seventies or the eighties, when tornado caves or dug- outs gradually came into use. At first there was some prejudice against these shelters, on the ground that it was undignified for a brave man to 'hunt his hole' at the sight of a dark cloud. But this idea disappeared, in time — and so did its proponents, if they tried to stay out in tornadoes.

Tornado destruction may start on the ground beneath the vortex forming at cloud level, even before the funnel cloud has appeared. People who have watched the near-by birth of a tornado tell of clouds racing along two or more opposite or conflicting paths, and then an inward, whirling motion of the clouds as the terrible vortex forms. Under the vortex trash and débris begin ascending spirally to the clouds. Then, in one reported instance, explosions began on the ground beneath: 'the ascending wreckage looked like the explosion of sparks from a great fire'; these explosions were 'similar to blasting operations.' After three separate ground explosions the funnel cloud formed, resembling 'a misshapen cornucopia,' disappeared, formed again, and moved away to the northeastward.

When a tornado is large and well developed, almost every ground object touched by the lower end or spout of the funnel cloud is carried away or destroyed. The path of total destruction, a sinuous track that may extend for many miles across country, is usually some three or four hundred yards wide. But if the funnel lifts, ground havoc ceases; and sometimes the funnel spout sails along perhaps thirty feet up in the air, tearing the upper stories from buildings whose lower stories remain unscathed.

Ordinary buildings caught in the funnel spout seem to explode, partly as a result of the sudden drop in pressure — roofs fly upward and walls outward. If windows are open, releasing the internal pressure, a building may be spared. Heavy bricks and concrete blocks mean nothing to the five hundred-odd-mile winds of the tornado —

they are strewn about like chaff. Modern, steel-frame buildings have apparently stood up fairly well, at least in smaller tornadoes. But not so with ordinary frame buildings. Trees are uprooted and laid flat or carried away; windfall tracks in forest country are graphic and mournful records of tornado passages in bygone days. Curious projectilelike effects occur in the fast-whirling winds — straws shot through boards, boards shot through logs and trees, and stones blasted through walls. Heavy timbers were carried completely away, in one instance, to varying distances between twelve and twenty miles. In another twister an eight-hundred-pound ice chest was carried several miles.

Another whirl whisked aloft a large bull and dropped him to earth forty yards away, unhurt, but frantically bellowing. A horse was carried half a mile without being killed; another was flung two hundred yards and cut in two. A child was carried half a mile and merely scratched. A thirteen-year-old boy flew three quarters of a mile to land again on solid ground, bruised, stunned, and stripped of his clothing. A woman was carried seven miles and killed. Tornado wounds, incidentally, are difficult to cleanse and heal, being deepiy scored with gravel and débris.

No vehicle can continue its course through a real tornado. Airplanes must avoid them at all costs. On a main highway a speeding sedan automobile, struck by a rather small twister, was lifted bodily, carried eighty-five feet, and dropped undamaged on its wheels in the opposite traffic lane. Some years ago the 'Empire Builder,' fast limited train of the Northwest, was struck by a sizeable tornado while rolling at a mile a minute. The tornado moved in from the side nearly as fast, struck the train fairly, and wrecked it. Five seventy-ton cars were lifted bodily from the rails, one being moved eighty feet. The onehundred and forty-ton locomotive and one hundred-ton tender remained on the track, but the cab windows were blown out and the engineer's goggles snatched from his head. The heavy steel cars saved most of the passengers, as in other train wrecks, from serious injury.

As for tornado protection, the lower inside rooms of a modern steel-frame office building are reasonably safe from twisters, as from all earth storms and earth shocks. There is also the old, reliable tornado cellar (alliteratively but incorrectly called 'cyclone cellar'), or dugout. After a dugout, the next safest refuge is probably the southwest corner of the cellar of a woodframe building (which will most likely be carried bodily away to the northeast). As in earthquakes, unframed buildings of brick and stone are often worse than none.

For people caught outside, the first requisite is quick but careful observation of the tornado's movement and probable path, together with an equally hurried orientation of compass directions. Most tornadoes advance out of the west or southwest towards the east or northeast. Plans for flight must also take account of the sinuous swaying back and forth of the funnel cloud, like that of a snake advancing on its victim. From directly in front of the tornado, it is best to run towards the north or northwest (as the northern side is likely to be slightly less violent). At any rate, never retreat to the eastward or northeastward unless we are in a vehicle capable of outracing the funnel. If flight is impractical, the best outside refuge is a deep, narrow trench or ditch, or the best hole or gully that comes to hand, or the eastern side of a large boulder mostly embedded in the ground. If there is nothing for it but an open field, lie face down, head to the east, and arms over head.

Seeing the Inside of a Tornado

As most human observers caught near a tornado are (to say the least) excited, and weather instruments caught within one are usually destroyed, full and accurate tornado observations are rare. By long odds the calmest and most penetrating tornado account that has come to my knowledge is the following one, written some years ago by Alonzo A. Justice for the *United States Monthly Weather Review* from the eyewitness story of Will Keller, a farmer of southern Kansas:

An umbrella-shaped cloud in the west and southwest and from its appearance suspected that there was a tornado in it. The air had that peculiar oppressiveness which nearly always precedes the coming of a tornado. But my attention being on other matters, I did not watch the approach of the cloud.

However, its nearness soon caused me to take another look at it. I saw at once that my suspicions were correct, for hanging from the greenish-black base of the cloud was not just *one* tornado, but *three*. One of the tornadoes was already perilously near and apparently headed directly for our place. I lost no time therefore in hurrying with my family to our cyclone cellar. The surrounding country is level and there was nothing to obstruct the view. There was little or no rain falling from the cloud. Two of the tornadoes were some distance away and looked to me like great ropes dangling from the clouds, but the near one was shaped more like a funnel with ragged clouds surrounding it. It appeared to be much larger and more energetic than the others and it occupied the central position of the cloud, the great cumulus dome being directly over it.

Around the lower rim of the great vortex small tornadoes were constantly forming and breaking away. These looked like tails as they writhed their way around the end of the funnel. It was these that made the hissing noise. The opening was entirely hollow except for something which could not exactly make out, but suppose that it was a detached wind cloud. This thing was in the centre and was moving up and down.

After it passed my place it again dipped and struck and demolished the house and barn of a farmer by the name of Evans. The Evans family, like ourselves, had been out looking over their hailed-out wheat and saw the tornado coming. Not having time to reach their cellar they took refuge under a small bluff that faced to the leeward of the approaching tornado. They lay down flat on the ground and caught hold of some plum bushes which fortunately grew within their reach. As it was, they felt themselves lifted from the ground. Mr. Evans said that he could see the wreckage of his house, among it being the cook stove, going round and round over his head. The eldest child, a girl of seventeen, being the most exposed, had her clothing completely torn off. But none of the family were hurt.

WATERSPOUTS

Waterspouts are sea or lake whirlwinds, and occur in many parts of the world. At sea off North America they are most numerous in the southern Pacific, across the Atlantic doldrums, and up along the warm waters of the Gulf Stream. These 'great horn spouts' can occur at any time of year, in warm weather or cold. They can also form at any time of the day or night, though most likely (as are sea thunderstorms) around dawn or around noon — at which times over-ocean lapse rates are steepest. There is great variety of form and intensity between different spouts, and in any single spout, even, at different stages of its life cycle. But by considering the way in which they are formed, we can divide all of them into two general classes — fair-weather spouts and storm spouts.

Similar in general mechanism to a dust whirl on land is the fair-weather waterspout. In place of desert intensive surface heating and absolutely unstable air, however, the sea affords a large supply of moisture which, by condensation, provides large quantities of latent heat. The over-ocean lapse rate may be moderate — say -3° or -4° F. per 1000 feet (about -6° or -7° C/km) — the air only conditionally unstable; yet the temperature and humidity are both high enough to insure that this conditional instability shall be realized in the form of updrafts and clouds.

Here again the whirl itself originates in a rising current fed by conflicting winds, and may materialize at the start as nothing more than one or two rings of spray on the water surface. As its feeder winds are neither very strong nor very extensive, and the earth's deflective force is weak near the equator, the fair-weather waterspout, like the dust devil, spins either clockwise or counter-clockwise by chance. True, a fairweather sea spout is usually somewhat bigger and more violent than a desert dust whirl. Yet in ferocity it falls far short of either a true tornado or a large storm waterspout formed in tornado fashion — that is, by the conflict of large and rapid and conflicting upper-air currents at cloud levels.

A waterspout at sea is always a fascinating sight that brings all hands on deck. While doing some marine weather work aboard the American freighter *Timber Rush* in April, 1939, it was my good fortune to observe several fair-weather spouts in about lat. 8° N., long. 84° W. — that is to say, about fifty sea miles off the south coast of Costa Rica. Records my journal:

At this time the sun had advanced well north of the equator, swinging three or four degrees northward of our zenith at noon; the rainy season was imminent; and the hot, moist, fitful weather belonged to that shifting equatorial belt of calms and sweats and squalls known as the 'doldrums.' Daily great white thunderheads towered upward. There came a day hotter than usual (eighty-five degrees or so), and more humid (dewpoint around eighty degrees), when the thunderheads towered white against the blue sky and black against the vertical sun.

Off the starboard bow, to leeward in the light breeze, there developed gradually a towering dark heap of cumulus cloud — a typical thunderhead. Its level but ragged base was perhaps about fifteen hundred feet above the ocean; its mushrooming top had reached perhaps ten thousand feet. Suddenly a narrow black funnel, a thin, horn-shaped cornucopia of cloud, grew downward from the cloud base. The ocean surface beneath the funnel erupted into an extraordinary fountain of whirling white spray as a long and whitish whirling tube, sinuous like the trunk of an enormous elephant, reached downward and bridged the gap between funnel and water. It was a waterspout, the ocean counterpart of the justly dreaded tornado.

The waterspout first formed about three miles to the northward of our course, and moved slowly towards us, traveling against the light surface breeze (in accordance with the similar drift of its parent cloud), even as we steamed ahead towards the track of the spout. There was no visible evidence, in the parent cloud, of pronounced, conflicting air currents; both surface and upper winds were light. The larger of these sea tornadoes are sudden death to any small craft they happen to strike, and have been known to strip lighter rigging and other loose top-hamper from a full-sized steamer.

The black cloud-cone, showing a lighter core, was now very plain; the lower, whitish tube writhed about on the ocean, picking up a fountain of white spray perhaps a hundred feet high and two hundred across.

But suddenly, at about a mile's distance, now almost abeam to starboard, the spout came apart in the middle; the whirling spray fountain quickly subsided; and the upper funnel very slowly drew up into the parent cloud above. Presently another funnel grew downward from the cloud, but never developed into a complete spout. Later, when the thunderhead was far astern, another spout grew downward, very long and very thin, and swayed slowly from side to side for several minutes before it broke up....

While an ordinary waterspout should do no real damage to a steel steamship, it is something to be avoided by any small-boat mariner. As a fair-weather spout usually moves rather slowly, and can be seen at long distances across the water, it is not at all hard to steer clear of them with a powered craft — except at night, or when, as rarely happens, a dozen or so may be wandering around at one time. Sailing-craft mariners without power, becalmed in front of an oncoming spout, can of course do nothing but make the best of whatever comes.

An account of a storm waterspout encountered in the seventeenth century appears in William Dampier's *Voyages*. The ship was of three hundred tons (larger than most yachts, fishing schooners and sailing coasters of the present day); yet it was manhandled and dismasted by a spout, while en route from London to the Guinea Coast in 1674:

When we came into latitude seven or eight degrees north, we saw several spouts one of which came directly towards the ship, and we having no wind to get out of the way of the spout, made ready to receive it by furling the sails. It came on very swift, and broke a little before it reached the ship, making a great noise, and raising the sea round it, as if a great house, or some such thing, had been cast into the sea.

The fury of the wind still lasted, and took the ship on the starboard bow with such violence, that it snapt off the boltsprit and foremast both at once, and blew the ship all along, ready to overset it; but the ship did presently right again, and the wind whirling round, took the ship a second time with the like fury as before, but on the contrary side, and was again like to overset her the other way. The mizzenmast felt the fury of the second blast, and was snapt short off, as the foremast and boltsprit had been before.

The mainmast and maintop-mast received no damage, for the fury of the wind (which was presently over) did not reach them. Three men were in the foretop when the foremast broke, and one on the boltsprit, and fell with them into the sea, but all were saved. The danger area for boats (or airplanes) around a waterspout is quite narrow — usually not more than a few hundred feet wide. In 1929 a United States Navy airplane flew within fifteen hundred feet of a small spout near Pensacola, Florida. Even at this distance the pilot noted no undue turbulence, and the observer was able to take a good photograph of the spout and its spray fountain.

Unlike the widely distributed waterspouts of tropic seas, temperate-zone storm spouts usually form along or in front of a marked cold front that is advancing eastward or southward over the sea, much as tornadoes form along a marked front on land. In tidewater regions, such as southeastern Virginia, a vigorous twister may change from tornado to waterspout and back again many times as it crosses bays and rivers.

Some years ago a whirl near Norfolk began by destroying trees and sheds on a point. Then it crossed a creek, sucking up the water so that the bottom was plainly visible and gouging out the exposed mud. Anchored small boats were hurled up onto the shore. Then it ripped off part of a heavy pier, destroyed some buildings, became a waterspout again in Hampton Roads, changed back to a tornado and dumped a railroad gondola car and some refrigerator cars off the tracks of a railroad yard, crossed and sucked up another creek, damaged some airplane hangars on land, and finally headed up Chesapeake Bay as a waterspout.

A storm waterspout may damage even a steel steamship. The *S.S. Hestia* passed through a fair-sized one, in 1902, at about sunset off Cape Hatteras. During heavy and widespread thunderstorms several spouts were seen forming, the largest of which headed directly for the ship in such a way that it could not be avoided.

The captain ordered all hands below decks, remaining out himself until the last moment. This oncoming spout was about fifty feet wide, with an almost black core about two feet across. As it struck, the captain dived below.

There was a deafening roar, strong wind gusts, and a sudden shock as the spout passed across the ship. Bobbing out again onto the bridge, the captain saw two great hatch tarpaulins and a large plank high overhead; and the log line from the taffrail, with the weighted spinner at its far end, extended straight up into the air like a large-sized Indian rope trick. In appearance a waterspout is quite similar to a tornado - the funnel itself is a hollow tube of spiraling cloud, formed by dynamic condensation. The usual spout height is one thousand to two thousand feet, though a specimen five thousand feet high was seen off Australia. Any large spout moves with the general drift of the cloud-level upper air — which upper drift may be across or against the surface wind. Around the base of the spout is the 'cascade,' a circular whirl of spray considerably wider than the spout itself.

A small spout cascade may be only fifty feet wide by ten feet high; but a storm-spout cascade runs larger than this — say two hundred feet wide by fifty feet or more high. Waterspout damage to ships is occasioned partly by tornadic winds, partly by suddenly reduced pressure, and partly by the deluge of water sometimes released. While an ordinary spout consists entirely of fresh-water droplets produced by dynamic pressure-fall condensation, there is good evidence that larger spouts carry considerable quantities of sea water up and away for some distance.

Probably the largest waterspout ever carefully observed by many intelligent people appeared between Martha's Vineyard and the Massachusetts mainland one sultry August day in 1896. Heavy, towering thunderheads were forming all about, and particularly banked up in the north and west, by noon of that oppressive day. A marked cold front (though unrecognized as such in those days) was advancing out of the northwest — it struck Cottage City on Martha's Vineyard later that afternoon, knocking the temperature from 72° down to 56°.

The great waterspout was first seen at 12.45 P.M., just as all the summer boarders were rising from a good old-fashioned dinner. The monstrous apparition formed and disappeared three times during the next half-hour. A long spiral column seemed to fall from the inky clouds, according to eye-witness accounts, increasing in size as it lowered and changing its colour from gray to black.

This record-breaking spout was seen by thousands of people (including a party on the yacht *Avalon* becalmed uncomfortably close to its cascade); and it was recorded by scores of photographers both amateur and professional — though not with the modern films and lenses and filters that would have captured a striking likeness. The ocean was flat calm, except immediately around the spout, as the immense black funnel slowly and majestically moved southeast toward Nantucket Sound.

The recorded dimensions of this great spout may strain credulity, but they were checked by too many reliable and scientificminded people to be lightly disregarded. The height was estimated at 3600 feet; the column was 840 feet wide at its top, 140 feet at its middle, and 240 feet at its base. The cascade was 720 feet across and 420 feet high.

Shortly after 1.20 P.M. the ponderous column drew slowly up into the clouds, and the aerological spectacle was ended. But salty rain fell on the Vineyard that afternoon, proving that this extraordinary spout had scooped up plenty of water from the Sound in the form of spray.

Chapter 10

Weather Forecasting

AEROLOGYS FUTURE

The future of any science can always be painted, what with the exponential rate of most human progress, as a brilliant picture in the most glowing colors. Of course this is particularly true of the great science of aerology, for its spacious laboratory, measurably colossal and immeasurably complicated, is the whole broad air blanket that is the Earth's atmosphere.

Any worker in the field of aerology comes across many factors of aerological equipment, technique, and procedure that are quite apparently in need of considerable improvement. To each intelligent and thoughtful worker there occur many possible solutions — 'inventions,' if we like — only a few of which he will ever have time to try out in actuality. What are the things that need improving in present-day aerology? What are the inventions, of equipment or method, that loom on the horizon of this, Earth's highest frontier?

Better Instruments and Observations

The most common aerological instruments are those devoted to measuring atmospheric pressure, temperature, and humidity. As to barometers, little if any future improvement over the present delicate and efficient instruments can be expected, and various difficulties combine to make any further advances along this line of doubtful value. We already have micro-barometers and micro-barographs which are more sensitive than we can practically use. An ultra-sensitive micro-barograph shows slight pressure variations occasioned by opening and shutting doors and windows, changes in wind directions or speed, earth tremors, and various other common natural occurrences which have nothing to do with the present or future state of the weather. Hence a less sensitive instrument is, for weather purposes, often more useful.

In the thermometer field, however, quite a few improvements can easily be envisioned. Perhaps the most obvious one is the more extensive use of remote-indicating thermometers of the electric type, in which the temperature-sensitive element can be located up to several hundred feet away from the indicating dial or recording pen. The widespread use of these electric gadgets, now entirely too expensive but potentially much cheaper if produced in mass, will permit better exposure of the sensitive element.

At present the thermometer shelters at many weather stations are located in untypical places (from a temperature viewpoint) such as valley bottoms, city districts, and so on. The result is that many local temperatures, as measured today, are inaccurate by several degrees. The same sort of improved technique might be applied to the measurement of atmospheric humidity — a characteristic fully as important as temperature in weather work. Here the key gadget might be a remote-indicating hygrometer of the electric-resistance type. Another technical development of possible usefulness might be a 'searchlight thermocouple' similar to those used by astronomers for determining the temperature of planets, stars, and so on. For aerological use these thermocouples might be pointed at clouds, portions of the upper air, and so on.

As indicated earlier in this book, the measurement of wind direction and speed at various places leaves much to be desired. The important wind for weather analysis, in most regions, is the 'gradient' wind or 'geostrophic' wind found at levels between five hundred and fifteen hundred feet above the ground or water surface. Tall radio towers and other high, open structures near weather stations should be equipped with remote-indicating anemometers wherever possible.

At larger radio stations, it might be worth building three hundred-foot steel-lattice towers just for temperature, humidity, and wind measurements above the ground-effect level. A promising gradient-wind-measuring gadget is a small mortar of about 1½-inch bore (similar to the Lyle gun used in alongshore lifesaving), which fires a small paper projectile straight upward. The projectile is filled with black powder and fused so as to explode at several hundred feet altitude. The resulting cloud of white smoke can be observed with optical instruments and its direction and speed of drift of course give the desired data on the gradient wind.

Another promising gadget is the cloud-thickness meter. In simplest form, this is merely a sensitive photometer (a good photographic exposure meter will do) pointed straight upward. With the sun at equal elevation, the observed light-intensity reading is of course roughly proportional (inversely) to the total thickness of clouds above the station — a bit of information often useful in weather work, particularly in aviation. Still another gadget is the visibility meter. Where visibility is the same thing as air transparency, a measurement can be made by some such method as having two white circles or other targets of unequal size, placed at such distances that they appear equal in size (or angular diameter) to the observer. The comparative clearness with which each device can be seen is then a measure of atmospheric transparency or visibility. Another recent development in the study of the upper air employs polarized light. An aerologist at Pennsylvania State College has found that, by measuring sky polarization with a special instrument, it is possible to sense the increase in size of condensation nuclei which usually precedes any sudden formation of lower clouds or fog.

For measuring ceiling (or height of cloud base) at night, vertical searchlights are in common present use; and whenever the ceiling is below a mile, and provided the searchlight is powerful and accurately focused, this method leaves little to be desired. In the daytime ceilings are now measured by means of small balloons that are supposed to rise at a fixed rate.

But this method has many disadvantages, and something better is badly needed. By using high-power radiation in some suitable portion of the visible or invisible spectrum, or by using polarized light, it might be possible to devise a sort of 'day ceiling light.' An- other possibility in day-ceiling indicators, which was suggested to me in 1937 by Mr. George A. Wolcott of the Weather Bureau, might be a rocket that rises vertically and leaves a visible trail of colored smoke. The point where this smoke trail intersects the cloud base, sighted on by an inclinometer at the other end of a base line, would of course give accurately the existing ceiling, provided the rocket actually rises vertically. Or perhaps a smoke-trail projectile could be fired vertically out of a smoothbore mortar.

Radio can be used in several fields of weather investigation. One that any amateur can try out is based on the fact that a lightning flash produces radio waves (mostly low-frequency around 200 kilocycles) that can be heard in any suitable receiver. Hence an increase of this 'crash static' usually means thunderstorm activity in the vicinity; and by using directional receivers along fixed base lines, thunderstorms can be located and followed in their wanderings. A somewhat similar radio-direction-finding technique can be used to locate and follow small tropical hurricanes (which usually have a good deal of thunderstorm activity near their centers) from fixed goniometric stations along the coast.

Short-wave radio can be used to some extent in locating and studying cloud masses in the upper air, the boundaries between air masses of different temperature or humidity, and other phenomena of importance in weather work. For example, in 1932. an amateur operating an ultra-short-wave (five-meter) portable transmitter on the summit of Pike's Peak found that clouds drifting between him and the other station on the plateau far below would cut out the radio signal entirely, apparently blocking it off by reason of refraction, reflection, absorption, or some other effect.

Work of this sort has also been carried on by two ultra-short-wave radio stations operating under the auspices of Harvard University on Mount Washington in New Hampshire and Whiteface Mountain in upstate New York. Since 1935 Mr. A. W. Friend of Harvard University has carried out extensive radio explorations of the troposphere with radio frequencies around 2400 kilocycles (about 125 meters). The apparatus can locate at will various discontinuity surfaces in the troposphere: temperature inversions (including the stratosphere base); the sloping planes of warm, cold, and upper fronts; tops and bottoms of cloud layers; turbulence layers. From a radio-echo sounding of existing turbulence levels aloft, Mr. Friend can even tell a departing aviator at what altitudes he is likely to find excessively rough and bumpy air.

Another radio development which fills a real need in weather work is the fixed robot radio weather station or 'ray-mete.' This consists of various meteorological instruments which measure and record temperature, humidity, wind, and so on. By means of a complicated gadget (similar to the ray-sondes to be described further along in this chapter) the observed weather data is transmitted by radio to a remote recording station which may be several hundred miles distant. The robot station, somewhat in the manner of an automatic lighthouse, can be left 'untouched by human hands' for six months at a time. Obviously the development of this technique towards full practical usefulness will permit the establishment of robot weather stations at many important but inaccessible locations — high mountainpeaks, selected points in arctic wastes and tropical deserts or jungles; ocean positions one or two hundred miles off our Atlantic and Pacific coasts.

Completer Analysis and Forecasting

One thing that impresses any scientific worker in these days — or, in fact, any intelligent and widely read layman — is the increasing mechanization and mathematication of all human science. Much of this is doubtless necessary in our present world, but some of it is perhaps not so necessary and a good deal of it is rather mournful to contemplate, particularly if we are the type of aerologist or layman who likes to look at clear skies and clouds and sunrises and moonsets and rainbows and halos for their own sake, as things of surpassing ethereal beauty, without any thought of the complex mathematical formulae which may enter into the precise laws governing them. Nevertheless mathematics has a very definite place in aerology or any other branch of physics, and any serious worker in these fields must travel the long and tedious road that runs from algebra through calculus.

One might, for example, write pages of near-poetic description in order to give a clear idea of what a thunderstorm is like. But a scientist might describe the same thunderstorm somewhat as follows:

$$C_p \log T - AR \log p + 0.623 \frac{re_m}{p^\tau} = K$$

$$AR \log \frac{p^o}{p^x} + \left(\frac{r+K}{p^x} - \frac{r}{p^o} \right) \frac{0.623\, e_m}{T} - \frac{K_q}{T} = 0$$

$$C_p \log T - AR \log p + \frac{0.623(r+K)e_m}{p^\tau} = K$$

And his succinct statement might be more useful, to all the bright young men studying at Massachusetts Tech. and California Tech. in preparation for work with the Weather Bureau, the Army and Navy, commercial airlines, and all the other weatherwatchful agencies in the modern American scene.

We have not often in this book bedeviled the lay reader with technical equations such as those above, nor will we in the future. In my own opinion there has been a tendency in recent years for mathematics to run away with physical science -something that often looks strangely like a small tail wagging a large dog. Mathematics is at best a shadowland of the mind which does not, and never will, represent concrete reality in any adequate and satisfying fashion. And while we are involved in this digression, it might be further said that the importance of physical science in general and certain aspects of technology in particular have perhaps been somewhat exaggerated. Far more important than all physical science, at least in the future, is *social science* — not how to invent more machines and methods, civil and military, but how to use intelligently and constructively, for general human benefit, those machines and methods that we already have in abundance.

A powerful new technique in weather analysis called 'isentropic analysis' was developed in 1937 at Massachusetts Tech. by Professor

Carl-Gustave Rossby and his associates. The general idea of this method, in the briefest terms, is as follows: Potential temperature usually increases with altitude. Hence it should be possible by means of many atmospheric soundings over a given region to construct a diagram showing the conformations of the various isentropic (equal-potential-temperature) surfaces above that region. Further, choosing one isentropic surface in middle altitudes it should be possible to draw a sort of contour map of it, in the same way that one draws a topographic contour map of some smaller region of the earth's surface.

In the troposphere, the potential temperature normally increases upward at a rate of about five Centigrade degrees per kilometre. It increases southward at about the same rate as the ordinary temperature. Thus the troposphere may be subdivided in a great number of thin isentropic strata (layers of constant potential temperature) which gradually descend towards the equator.

Isentropic-analysis charts are usually drawn and studied by aerologists in connection with the atmospheric cross-sections described elsewhere in this book. The practical end result of these isentropic charts is, in particular, a knowledge of where in the upper air great tongues of dry air are invading territory formerly occupied by moist air, and where similar tongues of moist air are penetrating into regions formerly dry.

One weather-forecasting possibility which has for years been a perennial dream among aerologists is the matter of foretelling weather many days or weeks or months or even years in advance. Quite a few amateurs have dabbled in this field with more or less popular success — success achieved mostly because human beings are rather gullible, by and large, with a tendency to remember spectacular successes and forget less colorful fail- ures. Some of the really scientific work in this field has been concerned with advances and recessions in the polar ice caps. Most of all, this seasonal weather variability probably depends on changes in the character and intensity of the all-powerful radiation from the sun — changes evidenced by sun spots, bright flares or faculae on the sun; by magnetic storms, auroras, and the like on earth. This fascinating field in solar-terrestrial research will be examined at some length in the next chapter.

A more immediate development in long-range forecasting is the endeavor to forecast weather for several days — or say up to one week — in advance. The Germans have done considerable work in this field, to which their plodding, methodical minds are ideally suited. Perhaps the most noteworthy work has been accomplished in the United States, under the auspices of the Weather Bureau — by Professor Hurd C.

Willett, of the Massachusetts Institute of Technology and his associates. In Willett's words:

It has been realized since the daily analysis of the Northern Hemisphere map was started at M.I.T. in the fall of 1936, that variations of the general circulation pattern, as defined by the intensity and position of the principal centers of action of the northern hemisphere, largely determine the persistent anomalies of the meteorological elements which are observed on this hemisphere.

In less technical language, the usual North American 'centers of action' directly control the advances and retreats and conflicts of air masses and fronts that cause our day-to-day weather. An unusual intensification of the Canadian-Alaskan high in winter, for example, usually causes a series of cold waves some days or weeks later over much of the United States. An extraordinary intensification of the Bermuda-Azores high in summer usually wafts a series of hot waves over the same region.

HIGH-PILOTS AND RAY-PILOTS

As early as 1935, it was apparent to some aerologists that long-range aviation was gradually moving upstairs. Stratosphere flying, or at least substrarosphere flying, was talked of as something bound to come within the next few years. The standard pilot balloon in ordinary use at that time, leaving the ground with a diameter of something like two feet and a free lift of about five ounces, usually disappeared from the observer's view before it had ascended more than four or five miles, even in the clearest weather; and at night, the small electric light or candle lantern which it carried aloft seldom permitted observations above altitudes of two or three miles. These altitudes, indeed, were likely to be considerably less if visibility were low or if upper winds were unusually high. Yet people were talking of flying the long-range, high-speed air transports of the future at altitudes of five or six miles; and frequent wind reports along the course ahead had become an essential part of every airline pilot's bag of navigating tricks.

On the Stratosphere Expedition of May-July, 1935, in South Dakota, described at some length in the opening chapter, we carried out some preliminary experiments in the direction of bigger and better pilot balloons and more powerful light sources for night use with these balloons. For the sake of simplicity, both these balloons and their night lights are here called 'highpilots.' In that clear North Dakota air, using two theodolite stations located about five miles apart at an altitude of about a mile above sea level, we got some remarkable observations extending to unusually great heights.

Continuing this 'high-pilot' research project back in the East during the next two years, some of us in the Army Aerological Service made hundreds of other balloon runs to high altitudes at Bolling Field in the District of Columbia and at Aberdeen Proving Ground in Maryland. At about this time, in 1936, there became available an improved large pilot balloon (inflatable at the ground to a diameter of nearly four feet and a free lift of more than two and a half pounds), manufactured by an entirely new process — a cheaper and lighter balloon than any that had ever before been available for this sort of work.

The immediate effect of the new lightness in the balloon itself was to make possible a rise rate considerably faster than had been possible with the smaller standard pilot balloons — or even with the large, heavy sounding balloons that we had used in North Dakota. (This faster rise rate of course extended the altitude of possible upper-wind observation; because the balloon had less time in which to blow away to great distances.) We were soon able, in fact, by inflating the new balloon to a free lift of about forty ounces, to achieve rise rates around 400 yd/min - just twice as fast as the rise rate of the smaller pilot balloon in general use at that time.

Whenever the daytime skies were clear and visibility was good we were able to follow these balloons to their ultimate bursting altitude somewhere between forty and fifty thousand feet — unless exceptionally high upper winds blew the balloon away to leeward unusually fast. Most of the time the balloon would reach bursting altitude at a distance of twenty or thirty miles from the main theodolite station. One afternoon we put a balloon run to thirty-six thousand feet on the airways teletype — to the consternation of the Washington operator, in whose estimation, apparently, such things just were not done, or at least should not be done in the existing state of the art. By hanging a five- or six-ounce weight on a ten-yard string beneath the balloon, it was possible (paradoxically enough) to speed up the rise rate still further — apparently because the balloon was pulled out slightly into a pear-like, or roughly streamlined, shape.

In scores of night balloon runs, trying for record-breaking altitudes under clear skies (clear for the East, that is, but nothing like as clear as Western skies in comparable weather), we tested various sorts of light sources — little acetylene burners supplied with gas from a smaller balloon hanging below the main lifting balloon; red pyrotechnic flares (similar to railroad danger signals) that were specially made up for us by a fireworks factory so as to burn with constantly increasing candlepower; three-watt electric lights consisting of a 1½-candlepower automobile bulb lit by six medium-sized flashlight cells; and a large tissue-paper lantern, cylindrical in shape and about one foot in diameter by three feet high. (Inside the lantern, attached to the bottom ring, was a compact group of four standard candles, all burning merrily away with a combined radiance of five or six candlepower. Both the acetylene light and the pyrotechnic flare never continued burning above twenty thousand feet — perhaps owing to oxygen shortage, or extreme low temperature.

The large tissue-paper candle lantern was, of course, a clumsy thing to carry around; and particularly difficult to launch, when the surface winds were high and gusty, without setting the whole lantern afire and destroying it then and there. (The lantern was completely closed with tissue paper at both top and bottom and, strangely enough, continued to burn brightly during the ascent with this small amount of ventilation.)

Nevertheless this old-fashioned candle lantern, once successfully launched, proved to be our best light source for dependably reaching high altitudes at night. Our highest Eastern night observation with the four-candle lantern extended to 25,500 feet altitude above station, checked by double theodolite on the twentieth minute when the lantern was about thirteen miles distant and encountering seventy-mile winds. Later we developed a candle unit in which four candles started the run, and were arranged by connecting wicks to light four more shorter candles after ten or fifteen minutes, giving twice the candlepower when the light was far distant and faint to the vision, but there was no opportunity to test this improved lantern in actual ascents before the whole project ended.

In many states having large forest areas it is illegal to send up candle lanterns or other burning devices with balloons. For this reason, and also to get a light source that could be launched without difficulty in high and gusty surface winds, we carried out many test night ascents with the three-watt electric light.

With seven or eight volts applied to the six-volt automobile bulb, it naturally burned blue-white with excessive brightness at the start; but as the balloon ascended and the minutes went on this brightness steadily diminished, and this diminution of the light itself, together with the increasing distance out as the balloon sailed away to leeward, made the traveling light speck among the stars harder and harder to follow as it gained increasing altitude. However, the electric-light source performed quite well on the whole. The best night altitude reached with it was 23,250 feet, when the light was about ten miles away in seventy-mile winds eighteen minutes after leaving the ground. A more powerful electric-light source, employing eight flashlight cells to give an effective starting voltage of ten volts or more on the six-volt bulb, and consequently a blinding light of 'photo-flood' intensity, was designed and tested on the ground.

Unfortunately, there was not time to make actual balloon ascents with this more powerful light before the entire research project had to be finished. When visibility is poor, or when fog or clouds obscure the sky, it is all too evident that no day balloons (however large) and no night balloon lights (however powerful) will serve to get the upper-wind reports that are often so important to aviators and aerologists. In dense haze infra-red radiation, rather than visible light, might help — but even infra-red is powerless to penetrate heavy clouds. For such work a still lower-frequency (or longer wave length) form of electro-magnetic radiation must be used — radio waves. And even the shorter radio waves may suffer absorption, reflection, or other distortion from clouds or fronts.

Hence the all-weather upper-wind-finder of the future, now scarcely beyond the laboratory stage, is the radio pilot balloon or 'ray-pilot.' While some earlier work had been done with impractical spark transmitters, the first important ray-pilot experiments with vacuum-tube transmitters were made about 1923 by Colonel W. R. Blair of the United States Army Signal Corps.

By about 1928 the Signal Corps Laboratories had further developed the ray-pilot into more effective and reliable form. The transmitter, weighing scarcely more than a pound, was surprisingly powerful and efficient. Many tests were carried out on wave lengths near 125 meters (2400 kilocycles), including careful radio-tracking of the balloon by means of specially developed direction-finding receivers on the ground, the radio bearings being checked by theodolite observations. It was found that the balloon could be easily followed, with useful azimuth readings, out to distances of ten miles or so from the ground stations. Within five miles' distance, the radio bearings were accurate to less than half a degree. No radio readings of the balloon's angular elevation were attempted with this equipment — its assumed altitude was figured from the average rise rate.

With the development of ultra-short-wave radio in the nine teen-thirties, it was apparent that this new frequency region offered many advantages for ray-pilots. The transmitters, as well as their antennae, could be smaller and lighter. The ultrashort-wave region was free from the high-power interference that cluttered up the longer-wave channels. Unlike the longer waves, ultra-short waves do not bend around the bulge of the earth or around hills or mountains, but follow a path, even as light does, that is straight or nearly straight. But this straightline propagation was no disadvantage in ray-pilot work — for the rapidly rising balloon would always be above any terrestrial hills or other obstacles.

Present in our 1935 Stratosphere Expedition were many distinguished scientists, and among the radio experimenters was Mr. Harry Diamond of the National Bureau of Standards. He soon joined forces with our pilot-balloon crew, and the result was a pioneering experiment in the field of ray-pilots and raysondes. We used no less than three large sounding balloons, all tied together in a loose cluster. Several feet beneath the balloons we hung a small radio transmitter operating at 108 megacycles (2.8 meters) with a radiated power of about one twentieth watt, and weighing about forty ounces complete. This weight was of course excessive, but the transmitter was rather hastily thrown together by Mr. Diamond and his assistant Mr. Lester from various available parts and batteries that happened to be in the Stratocamp.

This large balloon cluster was released from one of our baseend theodolite stations (though not without some difficulties occasioned by gusty surface winds) under a clear blue South Dakota sky at about noon. The little radio transmitter bobbed around erratically on the end of the string, and its signal consequently swung and faded badly. Also, the balloons went away to the northward in the lower levels, whereas the receiving beam antenna was pointed southeast.

Nevertheless, Mr. Diamond followed and measured the minute signal from the balloons at this ultra-high, and heretofore difficult, frequency for thirtyeight minutes on his superheterodyne receiver. At this time the balloon cluster was at an altitude of about forty-one thousand feet above sea level, and distant about twelve miles to the eastnortheast. After the radio signal died away the visual balloon run continued, followed by both our theodolites for over three hours. Between the eighty-seventh and eighty-eighth minutes one balloon burst, and the remaining two, weighted by the radio transmitter and the empty balloon, began to settle slowly downward.

The balloons had faded out to the naked eye after about half an hour; then, after three quarters of an hour, they had reappeared to the naked eye coming westward again in a marked east-west drift around 40,000 feet. At the eighty-seventh minute, the altitude of the balloons was about 60,100 feet above sea level, and the distance out about fifteen miles to the east-northeast.

At the one hundred and ninety-sixth minute, when the balloons were finally lost, they had descended to an altitude of about 15,000 feet, and were about twenty-six miles out to the east. The balloons and the radio transmitter were picked up about 3.30 P.M. of the same day by a farmer living about forty miles to the east.

After returning East to the Bureau of Standards, Diamond continued his pioneering work with ultra-high-frequency ray-pilots (as well as with the ray-sondes to be described in the next section). Using the remarkably high frequency of 200 megacycles (1½ meters), he carried out a variety of outdoor laboratorytype tests which clearly indicated the directions that future ray-sonde progress must take. The azimuth of a small ultrahigh-frequency radio transmitter drifting with the rising balloon can be found by two ground receiving stations at the ends of a suitable baseline.

Diamond determined, with an accuracy of something between one half and one degree. This accuracy is sufficient in the practical work of determining upper winds in or above the clouds. More important, the Bureau of Standards experiments showed that, by means of suitable antenna arrangements, the elevation angle of the ray-pilot could be determined by radio with an accuracy of one degree or less. Assuming that the ray-pilot was also acting as a ray-sonde — sending out a continuous record of the pressure it encountered, and hence its altitude — its distance away from the ground station could be readily computed. Thus the plotting of the invisible balloon's position in space could all be done from one ground receiving station, with even better accuracy than that possible where two azimuth-bearing stations were used.

Many technical difficulties are involved in this ultra-shortwave ray-pilot research, and the practical perfection of an efficient and reliable ray-pilot is still in the future. In most of Mr. Diamond's experiments so far, the transmitter has been raised by means of an elevator at a short distance from the receiver, rather than by a balloon rising freely in the atmosphere, in order that complete control over each step in the experiment could be maintained. Yet the technical difficulties are being gradually overcome, and before many years it should be possible for aerologists to know the upper winds by day or by night, come clear or come storm.

RISING RADIO ROBOTS

Visits to the stratosphere are to become a daily occurrence. Lieutenant W. H. Wenstrom of the Signal Corps, United States Army, has been making a study of radio sounding balloons at the California Institute of Technology and has indicated the feasibility of meteorological observations by the use of such instruments. At present the cost of production of a radio sounding balloon exceeds the cost of an airplane flight, but Lieutenant Wenstrom is very optimistic about the future use of radio sounding balloons. During bad weather when airplane flights cannot be undertaken, meteorological observations are most desirable. A radio balloon can be released in any weather and it will rise to much greater heights than could be negotiated by an airplane, with an instantaneous reception of the data on the ground.

At that time, when I wrote a graduate thesis on the subject, only four incomplete series of experiments with ray-sondes that actually ascended had been carried out (by Blair in the United States, Bureau in France, Duckert in Germany, and Moltchanoff in Russia); and although there was a great hullabaloo about large stratosphere balloons, the whole ray-sonde idea was *terra incognita* to most American scientists. as far back as 1875 by a Dutch instrument-maker named Olland.

In the briefest possible terms, the Olland telemeteorograph was as follows: Imagine a watch face, around which a sweep-second hand makes a complete revolution in some definite interval, such as every thirty seconds. Between twelve o'clock and four o'clock on the watch face is another hand, also pivoted at the centre; the position of this hand (over its own third of the dial) is regulated by the atmospheric pressure. Between four o'clock and eight o'clock is a third hand, similarly regulated by the atmospheric temperature. Between eight o'clock and twelve o'clock there is a fourth hand indicating the atmospheric humidity.

The thirty-second sweep-hand, driven by clockwork at uniform speed around the watch face, makes electric contact with the other hands in the following order: (1) a fixed reference at twelve o'clock; (2) the pressure hand; (3) the temperature hand; and (4) the humidity hand. Each electric contact would transmit a 'tick' signal by wire (or equally well by radio, which was still to come in Olland's day). The time intervals between the successive ticks, noted by ear or recorded by a suitable machine, would give a continuous record of the pressure, temperature, and humidity being experienced by the aerograph, whether it was installed on the ground (at a remote robot weather-reporting station), or rising through the troposphere with a sounding balloon.

In 1935, partly at the suggestion of some of us in the Army Aerological Service, Blue Hill Observatory of Harvard University started work on ray-sondes using the Olland telaerographic principle — the simplest, and in some ways the best, principle that will ever be applied to this work. Doctor K. O. Lange developed the aerograph part of the Blue Hill ray-sonde; Mr. A. E. Bent did the radio work. By the summer of 1936 this Harvard ray-sonde was emerging from the laboratory stage and had proved itself in scores of actual upper-air soundings. One of these, which I witnessed at Blue Hill in June, was reported to the Chief Signal Officer of the Army as follows:

In the Blue Hill instrument, the flat disk is replaced by a small insulating drum not much larger than a thimble. The pressure, etc. indicating arms move up and down on this drum like the pens of a standard recording instrument. The rapidly moving radial contact arm is replaced by a conducting silver thread wound in helical form on the small drum. Three quadrants of the cylinder are used respectively for the pressure, temperature, and humidity pens. The reference pen occupies the fourth quadrant, near the end of the drum; the helical silver thread makes one reference contact, and a small silver rivet makes another. This general principle and ingenious construction are believed to be the best so far developed anywhere.

The entire radio-sonde is enclosed in a cellophane casing, in order to absorb and retain solar heat at high altitudes. The radio part of the Blue Hill radio-sonde is also distinctive, in that ultrahigh frequencies are used. The very small, light transmitter operates on 68 megacycles (about 4.5 meters wave length) with an output of about one fourth watt.

As tested by the Weather Bureau in 1937-38, the Blue Hill raysonde weighed less than one pound complete, including the aerograph proper, the miniature radio transmitter, and its power supply. Also embodied in the improved Harvard instrument was a cheap, reliable, and reasonably cold-proof clock mechanism (weighing only two ounces) which drove the sweep-second hand (or rather, the small revolving drum) one revolution in thirty seconds. In order to adapt the instrument to night soundings, the 'green house' cellophane casing had been replaced by a heat-insulating balsa-wood box.

Another early form of ray-sonde in wide use by 1936 was the instrument devised by Professor Vaisala in Finland. Unlike the clock-driven aerograph element of the Harvard ray-sonde, Vaisala's radio-aerograph used as driving power a sort of small windmill, or arrangement of anemometer cups, which would steadily revolve in the downdraft caused by the steadily rising balloon. Another interesting feature of the Vaisala ray-sonde was a lead-acid-type storage battery of extremely light weight (three ounces) that would give constant voltage and could be recharged for repeated use. In the autumn hurricane season of 1936, an expedition from the Massachusetts institute of Technology used twenty Vaisala ray-sondes and ten Harvard raysondes for hurricane research in the West Indies.

Nineteen hundred and thirty-six was, in fact, a banner year in

American ray-sonde development. In response to a request from the United States Navy Department, Mr. Diamond and some of his associates in the Bureau of Standards were bringing to bear on the problem some heavy scientific artillery even as the first Harvard ray-sondes were rising into the New England skies. The specifications set up by the Navy Aerological Service were as follows:

The instrument was to have a unit cost of about twenty-five dollars in reasonable quantities, weigh about one pound, and provide pressure indications in the range of from 1000 to 200 millibars with an accuracy of indication of one millibar, temperature indications in the range of from +30 to -70° C. accurate to within one degree, and humidity indications in the range of from zero to 100 per cent relative humidity accurate to within 3 per cent. Consideration was to be given in the design to the provision of emitted signals suitable for use by radio direction finders on the ground in order to permit determination of upper-air wind conditions.

Mr. Diamond and Mr. Hinman and Mr. Dunmore did not like the idea of either a clockwork or an air fan to drive their aerograph mechanism. Hence they reverted to an earlier system which had been used by Duckert some years before — in which the pressure-indicating element itself served as the driving mechanism. As the balloon rises, the pressure of course decreases, steadily and more or less uniformly; thus the moving arm of the pressure element can be made to serve as an electric switch to close successive contacts, emitting characteristic radio signals. In the words of Diamond, Hinman, and Dunmore:

The pressure arm, moving continuously in one direction as the balloon ascends, switches the frequency of the audio oscillator to correspond alternately to the values of the temperature and of the humidity encountered. The alternate change-overs from one set of frequencies to the other indicate that the pressure arm is just reaching or is just leaving one of the intermediate contacts and his attained definite deflection positions. When the pressure arm reaches successive fifth conducting segments, the frequency of the audio oscillator attains predetermined fixed values which positively identify these contacts so that they may serve as index marks for the absolute-pressure scale.

At the ground station was a complicated radio receiver-recorder which would automatically plot a complete adiabatic diagram as the balloon ascended — temperature and humidity plotted as abscissas against pressure as ordinates.

In complete form the Diamond-Hinman ray-sonde, as manufactured by Friez of Baltimore and adopted for standard radiosounding use throughout the United States and its territories and its surrounding oceans by the United States Weather Bureau and the Navy Department in 1939-40, weighed about two pounds. The radio frequency employed as a carrier wave was about 65 megacycles (about 4.6 meters); the various modulation frequencies which indicated changes in temperature and humidity during the ascent were between 10 and 190 cycles, so that the radio signal from the balloon sounded somewhat like an organist practicing various low notes.

Below the balloon itself, on a ten-yard length of heavy twine, hung a small silk parachute, a springy shock-absorber ring, and the radio-aerograph itself, with the short transmitting antenna hanging downward from the aluminum-foil-covered, corrugated-pasteboard box. Usually the balloon would burst at about fifteen miles' altitude, the parachute would open, and the aerograph would descend safely to earth. On the average, about half of the ray-sondes released in settled regions are picked up and returned (in response to a reward of several dollars); and about two thirds of these, or one third of those released, can be repaired and used again. At sea or in remote regions, a released ray-sonde is of course lost for good; but the important upper-air data are nevertheless secured.

By 1940 'RAOBS' (ray-sonde observations) were being made nightly from something like forty stations scattered all over the United States, not to mention about ten ships of the Navy in various adjacent waters. In the Southwest, for example, radio soundings were being made at San Diego and Oakland in California, Phoenix in Arizona, and Ely in Nevada. Nevertheless the older method of airplane observations or 'APOBS' had not been entirely discontinued. Useful in many military and naval situations, the APOBS were still being made at some Army and Navy stations.

Still another American ray-sonde has been developed by Doctor L. F. Curtiss of the Bureau of Standards, using ultrashort-wave radio together with a ray-mete based on the original Olland telemeteorographic principle. Doctor Curtiss has also pioneered, along with such renowned American scientists as Karl Compton and Andrew Millikan, radio-sounding-balloon ascents far up into the stratosphere with the object of measuring cosmic radiation at these extreme altitudes.

One of the Curtiss ray-sondes, ascending from Washington, D.C., was followed by a Navy airplane from the Anacostia Station. The balloon of course rose far above the airplane's utmost ceiling — to a height of about seventeen miles; but as the burst bubble plummeted downward, the watchful Navy airman got close to it about two miles up and by diving steeply followed it down to the ground, where it was recovered, about thirty miles from Washington, by a surprised Maryland farmer. One of the Curtiss ray-sondes reached an altitude of about twenty-five miles or 132,000 feet above sea level — probably a world's record for any sort of balloon.

Still other Bureau of Standards scientists — Coblentz and Stair — have used ray-sondes for measuring the intensity of ultra-violet radiation from the sun in the stratosphere. Their instruments employed a photoelectric ultra-violet-meter connected so as to operate a miniature radio transmitter of the Diamond-Hinman type. Several ascents were made, which showed high temperatures and lethally intense ultra-violet radiation in the upper stratosphere. The importance of this research in such fields as long-range weather forecasting can scarcely be exaggerated, for the solar radiation which varies most widely and which probably affects earthly weather most violently is precisely in this ultra-violet region of the spectrum.

Another possible radio-robot device for sounding the blue depths above might be a sort of oversized model airplane — say with a wingspread of something like ten or fifteen feet and, horsepower of fifty or so — designed for rapid climb, radiocontrolled, and equipped with a ray-mete that would send back the essential p.,t., and f. data as the pilotless plane sounded upward. Many little planes of this general type are now in use as anti-aircraft artillery targets in progressive armies.

A further development in radio-controlled and radio-reporting gadgets for sounding the troposphere might use the helicopter principle. Professor von Karman of California Tech. (one of the world's authorities in aeronautics) suggested to me in 1934 that the idea should be feasible. The actual sounding device might be an oversized model helicopter, equipped with two large, oppositely revolving lift propellers on the same shaft, and balloonlike shock-absorber feet to cushion the landing jolt on descent. With efficient helicopter design, one horsepower will lift a total weight of something like fourteen kilograms (about thirty pounds). Hence for moderate-altitude sounding in all sorts of weather, one might envision a little whirligig six or eight feet in diameter, run by a miniature gasoline motor of about one horsepower and weighing complete (structure, motor, gas supply, radio control, and aerological ray-mete) about thirty pounds. The little spinner could take off vertically from the roof of the weather station, climb at 600 to 1000 ft/min straight upward under a ground aerologist's control, and perhaps return shortly, still under control, to a bouncy landing on the airport.

Searchlight Soundings

Sounding airplanes and helicopters, piloted or pilotless, may penetrate the stratosphere in long and arduous climbs to heights of eleven or twelve miles. Beyond this, conventional wings and propellers will not go. Expanding, hydrogen- or helium-filled balloons of thin rubber, rising in effortless fashion to their own destruction, may carry various instruments still farther aloft -to heights around twenty-five miles. Beyond this, rubber and hydrogen will not go. Yet even twenty-five miles aloft is merely the middle of the stratosphere, and only halfway to the beginning of the ionosphere.

The ionosphere can be probed, and to some considerable extent explored by intangible radio waves. Further facts about the upper air may be revealed by another intangible sounding method — modulated light. In briefest terms, this method (first suggested by M. A. Tuve and E. O. Johnson of the American Carnegie Institution and O. R. Wulf of the United States Weather Bureau) consists of (1) a searchlight on the ground that throws a beam of distinctively flickering (recurrently modulated) light into the upper air, and (2) a directional photometer (a photocell at the focus of a large parobolic reflector) on the ground that measures the reflective scattering of this searchlight beam in some chosen segment of the upper air.

The intensity of this reflective scattering, for white light and for various spectral colors, gives some very definite clues to the density of the upper air — 'a basic unknown in present-day upper-air studies.' Thus this light-sounding method may help to disprove or confirm some the fabric of careful scientific guesswork which constitutes our present knowledge of the upper atmosphere.

If the sounding searchlight projected a steady beam, its airreflected light would of course be indistinguishable from various interfering luminescences of much higher intensity — starlight, skylight, and so on. But by using a revolving chopper on the searchlight to interrupt or modulate its beam (at say one or more cycles per second), a sort of 'identification tag' is placed on the particular light being used in the sounding; and for greater sensitivity, the electric amplifier at the receiver can be designed to magnify this chosen frequency at the expense of all others.

In practice the searchlight itself can be a standard thirty-sixinch, ten-kilowatt anti-aircraft light, or a larger instrument such as the sixty-inch seacoast-artillery light. The parabolic reflector at the receiver can be identical with the searchlight reflector Both transmitter and receiver are pointed upward (at an angle of say sixty degrees) towards the upper-air segment under test; and they are of course located on the ground several miles apart, at an exact distance governed by the height of the light sounding.

Needless to say, crystal-clear weather and a moonless night are best for this delicate method of searchlight sounding, and the American regions most ideally suited to it are perhaps the high desert plateaus of Arizona and New Mexico. With more powerful transmitters (such as the new water-cooled, concentrated-mercury-arc searchlight), and with more sensitive and selective receivers, the method may yet solve some important enigmas of the lower ionosphere.

Rockets to the Fore

About the only known device capable of actually carrying safely an air trap, a ray-mete, a cosmic-ray recorder, an ultra- violet-meter, or other delicate instruments to heights above fifty miles is the rocket. This engine, alone in the whole field of technology, suggests once again the fascinating possibility of sounding, or perhaps even personal exploration, into the remoter depths beyond all possible reach of the airplane or the balloon. Depths these are, such as no man has ever gazed on, where the blue and violet have faded into brilliant darkness, black depths sunlit and starlit at the same time!

On a rare occasion there is vouchsafed to early rising humans a cosmic-terrestrial spectacle that gives a sort of preview of things likely to come — a sort of reverse rocket sounding at the incredible speed of forty miles per second — a vertically falling meteor or meteorite. One clear spring morning, moonless and crisp. Most meteors fall swiftly and brightly at this hour of the morning, but few of them fall vertically, and still fewer of these are bright enough to leave trails of glowing cosmic dust in the sky.

Suddenly (it was half an hour before sunrise) an extraordinarily bright meteor appeared perhaps sixty degrees up in the western sky. It plunged vertically downward, straight as a plummet and unbelievably swift, brightening as it plunged into a great rocket-star that cast shadows from the trees around me. Within a second's time the heavenly apparition flickered out at about thirty degrees elevation, and the cosmic spectacle was over.

Yet I stood rooted to the spot for many seconds more as the dim-glowing white trail slowly faded from view, awed by this unearthly visitor from outer space. Rockets have for generations been used in both peace and war. (The gruff Wellington commanded British rocketeers to lay by their cherished weapons in favour of more reliable guns at Waterloo; and in the 'rockets' red glare" Francis Scott Key conceived our National Anthem.) But the first thorough and modern scientific study of this miracle-machine and its possibilities was made by Professor Robert H. Goddard in the United States.

As a result of research begun at Princeton University in 1912, Goddard submitted to the Smithsonian Institution in 1916 a paper, entitled *'A Method of Reaching Extreme Altitudes'* -a treatise which has served as the basis for subsequent rocket researches all over the world. In laboratory experiments with smokeless powder exploded in heavy steel chambers Goddard bettered the performance of Fourth-of-July rockets eightfold in gas velocity and thirtyfold in efficiency. After 1920 he worked for several years on the production of a safe and controllable liquid fuel.

In other parts of the world, notably in Germany, rocket experimenters have also been active. Some few have lost their lives in the work, caught by the premature explosion of their powerful machines. Besides the pioneer work of Goddard, contributions of note have also been made by such experimenters as Herman Oberth of Rumania and Robert Esnault-Pelterie of France. On the outer fringes of rocketry, however, have swarmed a multitude of crank 'inventors,' many of whom were simply playing with Fourth-of-July contraptions on a glorified scale.

Old-fashioned black powder (a mixture of sulphur, potassium nitrate, and charcoal) does well enough for celebration fireworks, where it is only necessary to push colored lights up a few hundred feet. But for the task of pushing instruments up a hundred miles, it is about as useful as cordwood to an airplane; it simply does not contain the necessary energy. The only possible answer is liquid fuel — a mixture of liquid oxygen and some substance having an explosive affinity for it, such as liquid hydrogen or gasoline. So far most liquid-fuel experimenters have worked with gasoline, which is cheap and universally obtainable, and actually embodies ten times the total explosive energy of the most powerful military explosives. Combining with liquid oxygen in a combustion chamber, it develops a temperature of about 3500° F. or 2000° C.

The hear energy developed in the combustion chamber or 'rocket motor' blasts expanding gases out of a long, slightly flaring nozzle; and the rocket moves forward not by pushing against the air, but by throwing rearward at great speed a small weight of gas. Hence it should be perfectly at home in the partial vacuums of great heights above the earth's surface.

But for reasonable thermal efficiency (which may be as high as ninety per cent) the combustion chamber must be expertly designed; and the ballistic efficiency, though one hundred per cent if the rocket's forward speed is equal to the backward velocity of the exhaust, falls very low at speeds much less than the pistolbullet velocity of seven hundred miles an hour or one thousand feet per second. Controlled burning of the liquid fuel, also, presents a very ticklish problem. The liquid oxygen supply will violently explode unless its temperature is held far below zero -yet a few feet away, in the combustion chamber, is a raging furnace that would melt iron!

Some of the energy values obtainable from efficiently burning gasoline are startling, — at least in theory. If it could all be used with one hundred per cent efficiency in quick combustion with some independently available oxygen supply such as the free air, a single gallon of gasoline would be sufficient to blow an ordinary passenger automobile something like ten miles high! The same gallon might theoretically serve to lift a twentypound sounding rocket some fifteen hundred miles above the earth's surface.

But in actual rocket practice, each pound of gasoline must be accompanied by three and a half pounds of liquid oxygen — and there go our theoretical figures galleywest! *Actually, a hundred-pound rocket may contain ten pounds (one gallon) of gasoline and thirty-five pounds (three and one half gallons) of liquid oxygen, and might possibly attain a height of two or three hundred miles.* Such a rocket might possibly attain, in the extremely tenuous 'air' of the ionosphere (for such speed in the lower air would soon heat it up unbearably), *a speed of one or two miles per second.* (To escape from the earth's gravitation, a 'space-rocket' would have to attain a speed of seven miles per second — something that, unlike the one-mile-per-second sounding rocket, may here be left to the limbo of dreams-of-theultimate-future.)

Goddard's first liquid-fuel rocket flight was made in March, 1926. But the first flight that awakened widespread public at- tention was made on a clear summer day in July, 1919, from a farm near Worcester, Massachusetts. With a single exception, everything went according to schedule; the rocket was expected to rise only a few hundred feet and soared neither higher nor lower. The exception was public curiosity and consternation. Any high-velocity rocket, by its very nature, must make a terrific and eerie noise — a sort of continuous but explosive howling roar, perhaps comparable in nature only to the mightier roar of the tornado.

Scarcely had the last echoes of the thunderous 'whoo-oosh' occasioned by the world's first practicable liquid-fuel rocket died away, when the astonished Massachusetts countryside awoke to frenzied activity. Multitudes of automobiles converged on Goddard's Rocket Farm, and the retiring scientist, who never has liked publicity, was soon confronted by tourists, reporters, and even ambulances!

In recent years Goddard's experiments have continued on a remote ranch in New Mexico — a mile nearer the stratosphere and many miles farther from the mob. From this land of yellow sand floors, broken by sage and cactus, there have issued no bulletins of progress and no speculations about Mars or Venus. Nevertheless, work of the utmost scientific value has been steadily going forward.

On one occasion a large liquid-fuel rocket (a foot in diameter, eighteen feet long, and weighing more than eighty pounds) sounded to a height of about seventy-five hundred feet and descended safely to earth by parachute. Its marvelously efficient 'rocket motor' or combustion chamber, developing more than two hundred horsepower per pound of motor weight, quickly drove it upward along the sixty-foot guide tower and into the air.

Tipping slightly from side to side under the action of its gyroscopic stabilizer, it resembled nothing so much as a great fire-belching fish swimming upward; slowly at first, then faster and faster to the accompaniment of a steady and monstrous roar. A small white flame streamed backward from the tail of the rocket, occasionally flashing out in irregular gasoline explosions, and ending in a sinuous trail of black smoke. Before reaching its maximum altitude this sounding rocket attained a speed of seven hundred miles per hour — it was just beginning, in other words, to operate at high ballistic efficiency when the fuel ran out.

In reply to a recent letter of mine, Professor Goddard wrote of his work in 1940:

Two of the three developments involved in making highaltitude sounding rockets have been completed. These are the production of a satisfactory chamber, or 'rocket motor,' and the automatic stabilization of the rockets in flight.

Some day, perhaps, the desert may echo to a roar which startles the little horned toads and sends them scuttling under the dubious shade of mesquite and greasewood, while a pointed silver cylinder, streaming backward a fiery trail, rushes high and still higher into the turquoise New Mexico sky until it is lost to view, only to reappear, some minutes later, as the white puffball of a parachute bearing gently to earth the first actual sounding records of the ionosphere. And some day, not far distant, perhaps, long-range bomb-rockets may carry the avenger's message from the encircling armies of aroused democracy to Berlin and the other citadels of those aggressors who would live by the sword at the expense of all free and decent men.

THE EMPYREAN

Everyone knows that man inhabits an eight-thousand-mile ball which moves ever silently through space in accordance with fundamental law; less generally realized, though equally true, is the fact that man's locomotion is confined to a twenty-mile shell on the surface of the earth. All his express trains, all his ocean liners, all his racing airplanes, have brought him no farther.

Steam shovels, dynamite, and submarines have never opened for him depths below a mile or two. Balloons and climbing wings have never carried him to heights above fifteen miles. But after sounding rockets have risen high and floated safely back to earth many times, demonstrating the reasonable safety of the method — perhaps long after these pilotless robots have broken the upper frigidity — may come the hour for man himself to escape his ancient bondage.

What splendors the first explorers will see! Cloud levels flash by like floors past an express elevator, to be succeeded by the deep cloudless blue of the lower stratosphere. As the sunlit rocket sounds to greater and still greater depths the blue deepens into violet or purple, 'with stars beginning to appear. The purple in turn darkens, and against the blackness of empty space planets and bright stars shine out with steady brilliance like enormous and remote lamps.

Magnificent are the star streams and clusters beyond any incomplete earthly vision, and among them hang immense nebulous clouds proclaiming in diffuse light the eternal message of fire-creation. Against all this infinite pageant the great sun glares blue-white, a thousandfold more fierce than accustomed daylight, yet fringed by the pearly aureole of the corona. And far beneath is the broad earth, expanded to new horizons, hazewrapped and cloud-littered; yet visibly bulging, evidently a part of the greater cosmic scheme, viewed for the first time from the mythical slopes of Olympus.

Chapter 11

Solar Flares and Earthly Storms

Despite our claim as its very children, the Sun must acknowledge even closer kinship with the widespread galaxy of suns that we call the stars. Out in this far-flung galaxy, and beyond its limits in other galaxies more remote, strange things happen.

Some of those distant suns, far bigger than ours, pulsate regularly every few days like the slow beating of immense fiery hearts in the depths of space. Other suns, which have been shining steadily in mediocre fashion for thousands of years, suddenly flame out in prodigious brilliance.

On the fifth of June in the year nineteen hundred and eighteen, for example, Nova Aquilae appeared to men as an ordinary telescopic star, too faint to be seen with the naked eye. On June seventh it became apparent that the furnace fires of Nova Aquilae were running wild. Hourly the star was getting brighter. At the peak of the explosion it was almost the brightest fixed star in the sky, rivaling even Sirius, and its radiation had increased about fifty thousand times.

If our sun suddenly decided to become a nova, what disastrous earthly effects would ensue! Radio messages telling of fierce and then scorching heat on the daylight side of the earth. Actual fire in the tropics; arctic ice melting in cascades. Then charred silence. Sunrise a wall of flame, searing every combustible thing to blackened ashes.

Our Sun as a constant star — an unchanging source of light and heat. This chapter deals with solar changes. Fortunately for us the sun does not pulsate in easily visible degree, and its chances of becoming a nova are exceedingly small.

But we cannot take it for granted that the Sun will never vary, nor can we say with absolute certainty that some catastrophic change in it will not occur tomorrow. Our Sun is actually a variable star, changing in rhythmic fashion through a complex but repeating harmony of periods of time. And the vibration chords that our Sun sounds in heaven awaken on this earth echoes that profoundly influence many things ranging from telegraphy through radio and aurorae and magnetic storms to weather.

Sun Spots and Hot Spots

The most plainly apparent of solar variations is the cycle of the sun spots. Sun spots appear through the telescope, or through good binoculars when large, as dark spots on the photosphere. Most of them are roughly circular in shape, and they tend to occur in groups. The centre of a spot (called the umbra) is darker than the outer ring (penumbra). A sun spot usually begins as a small black point which grows into the umbra; then the penumbra fills out. The spot appears to travel slowly from left to right on the sun's disk, of course, with the solar rotation. After a few days, or more likely after a few weeks, the spot slowly decreases in size and finally disappears. As the temperature inside a sun spot is about 4000° F. (about 2000° C.) below that of the surrounding atmosphere, it may be called the greatest refrigerator known to man! Though smaller spots are only a few hundred miles in diameter, larger spots or groups may be fifty thousand miles wide — holes into which the whole earth could be dropped like a golf ball into a hat. These larger spots are visible to the unaided (but protected) eye.

Naked-eye sun spots were noticed many centuries ago by the Chinese, who regarded them with vague fear. The first man to examine sun spots in some detail through a telescope was Galileo, in 1610. The first man to study them seriously was Samuel Heinrich Schwabe of Dessau, apothecary by trade but astronomer by avocation. His pills, like Keats's, are long since forgotten; but his patient observations through the years un- questionably established *the eleven-odd year cycle in the increase and decrease of sun spots*. This cycle stands today as one of the basic facts in solar variation.

With each new cycle, sun spots begin to break out about 36° north and south of the sun's equator. In a few years, with maximum activity, most of the spots are much nearer the equator; and the old cycle finally dies out between 5° and 10° north and south as the next cycle is getting under way in higher latitudes. The sun bursts into activity with comparative suddenness — in about four years; and simmers down more slowly through about seven years. The last sun-spot maximum was in 1937-38; the next is scheduled for about 1948, with a minimum around 1944, In addition to the eleven-year periodicity of sun spots, overtones of from six to fifteen months appear to be superimposed on the larger and slower fundamental. Also, of course, there is a twenty-seven-day period imposed by the sun's rotation.

In the year 1908, George Ellery Hale made in California a

significant discovery that revealed the true nature of sun spots. Ordinary visual observations, it is true, had occasionally shown a vortical or whirling motion of gases in the neighborhood of the spots. When photographic plates sensitive to red light became available, Hale photographed the sun in the light of one spectral colour only. He was enabled to exclude all other colors from his photographs by means of an ingenious instrument known as the spectroheliograph. As the single colour he chose was emitted by a single gas, hydrogen, he secured in effect a picture of the sun's surface at one particular level — the level where hydrogen gas predominated. The photographs plainly showed gigantic whirling currents about the sun spots, similar in principle to earthly cyclones. And like earthly cyclones, the solar vortices whirl in opposite directions on opposite hemispheres.

Hale showed further, by intricate light measurements, that a sun spot is a great electro-magnet having an intense magnetic field, which is probably induced by the free electrons and ions in its rotating gases. A sun-spot magnetic field probably extends outward in a more or less directed way like the field of a bar magnet; and interferes, locally, with the main magnetic field of the sun (which in itself is much like the earth's field that guides all our compasses, but about one hundred times as strong). Hale found also that the sun spots change magnetic polarity from one cycle to another, so that the complete magnetic cycle of the sun averages twice eleven-odd years, or twentytwo to twenty-three years.

The occasional observation of sun spots is an avocational activity of interest to any intelligent person; and particularly so, in view of correlation possibilities, to anyone interested in the weather. Very little apparatus is needed for a start. Even the poorest sort of opera glass, field glass, or telescope will reveal spots when solar activity is high. It is necessary, of course, to use some sort of light-absorbing screen between the sun and the eye — perhaps a piece of glass blackened over a candle, or better, a thickness or two of overexposed photographic film. Some of the better binoculars and telescopes can be obtained with eyepiece filters of dark glass, which slip on conveniently and cut down the excessive light. By turning the eyepiece of a large hand-glass or a small telescope to a suitable focus, the solar image (spots and all) can be projected on a shaded piece of white paper and traced over in pencil.

Often seen near spot groups on the sun are bright 'faculae' (named from the Latin word for 'little torch'), which appear as bright, lacy patterns on the solar surface. It is not possible to see them at all without a fairly good telescope or a very excellent hand-glass; and even with one, they are visible only near the sun's edge, or limb, where the background of solar surface appears darker. The faculae are supposed to be ridges or elevations in the photosphere, and it seems quite reasonable to assume that they are *'hot spots' in relation to most of the solar surface -areas which radiate more than their share of total energy, and much more than their share of powerful ultra-violet.*

In addition to the faculae, there are bright clouds of calcium called 'flocculi' (Latin 'tuft of wool'); and also dark flocculi, which are clouds of calcium and hydrogen at high levels in the chromosphere. These dark flocculi, incidentally, are sometimes observed being drawn into the upper levels of sun-spot vortices.

While the dark and easily seen sun spots have been assiduously observed and studied for generations, all too little intelligent human attention his been directed towards the faculae or solar flares — the solar hot spots that bombard the earth's outer atmosphere with monstrous intensities of ultra-violet radiation, and colossal quantities of high-speed free electrons and protons.

Scientists without number have tried to correlate magnetic storms, auroras, radio reception, and even weather with the day-to-day or year-to-year rise and fall of the sunspot numbers — and with striking success in some fields. But the success might be even more striking, particularly in the most difficult and complex field of all solar-terrestrial correlation: world weather — if researchers paid less attention to sun-spot numbers and areas; and more attention to favulae areas and brightness indices.

Usually the rose-tinted chromosphere of the sun is blotted out by light from the blinding photosphere. But during a total eclipse, or viewed through the spectrohelioscope, its ruddy ring becomes visible. It is then seen to be very irregular at its outer edge, with great flame-like prominences often rising a hundred thousand miles or more above the solar surface. Quiescent prominences often keep their general form for days on end, and are probably caused by the luminous excitation of diffused, highlying clouds of helium, calcium, and hydrogen. But prominences may become eruptive, and are then gigantic explosions wherein the metallic vapour travels outward at a hundred or more miles a second, reaching heights of two or three hundred thousand miles. These great solar flames dwarf the largest earthly volcano into the comparative insignificance of a toy spitfire.

Both faculae, or hot spots, and prominences are likely to be more numerous and more intense in the neighborhood of large sun-spot groups than elsewhere on the sun. Also, faculae-andprominence activity tends to wax and wane with the sun-spot cycle; the solar eruptions are more prevalent in years of sun-spot maximum. Faculae and prominences, indeed, are probably parts of the same thing viewed from different angles.

The 'solar constant,' or measure of the sun's total radiation, has been measured by complex and ingenious methods devised mainly by Doctor C. G. Abbot, solar researcher of the Smithsonian Institution in Washington. Measured on the earth's surface, this entity would be meaningless because of variable atmospheric influences. But the Langley spectro-bolometer and the Abbot pyrheliometer permit aerologist-astronomers reliable to determine the solar constant outside the earth's atmosphere from terrestrial observations at moderate altitudes. By this method, checked and rechecked many times, the average value of the solar constant, outside the earth's atmosphere at mean earth-sun distance, is found to be about two calories (1.94, to be exact) per square centimeter per minute. That is to say, the solar radiation falling squarely on a centimeter (.6 inch) cube of water would, if entirely absorbed, raise the water temperature about 2° C. (around 3° F.) per minute.

From the mean of the solar constant values reported by its three mountain-top stations, the Smithsonian Institution is enabled to plot with confidence a curve which represents the actual day-to-day, month-to-month, and year-to-year variations in the solar constant. At first glance this curve resembles nothing so much as a profile of the Alps, and seems to show no regular periodicity. But a complicated machine is available which can analyse such a complex curve; it can, for example, break up the complex and irregular curve representing the playing of an organ into the pure and regular vibrations representing the fundamental notes that go to make up a chord, and all their separate overtones. In the same way, this machine analyzes the complex output curve of the solar light-organ; and finds that, in addition to the fundamental periods measured in years, there are many overtones measured in months. And there is, of course, the twenty-seven-day period imposed by the sun's rotation.

The changes in the constant of total solar radiation are small, seldom exceeding three per cent. And it is fortunate for us earthlings that they are; for if the constant changed by as much as fifty per cent, life would vanish from the face of the earth.

Of all the electro-magnetic energy wave lengths radiated by the sun, the shorter wave lengths vary the most in response to solar disturbances. Though only a small fraction of the total radiation is in the ultra-violet region, this region suffers the greatest changes with the changing solar constant, and may mainly account for some of the earthly echoes that respond to the solar chords. Around 1932. Edison

Pettit of Mount Wilson Observatory devised an instrument which measures the ratio of the sun's ultra-violet to its visible light. As a sample of visible light Pettit chose green, in the centre of the visible spectrum. To get green and no other colour one might use a filter such as those used in three-colour photography — perhaps a piece of green glass. But Pettit wanted a narrower band of green than most people (at about 5300 a.), so he chose for his filter a film of pure gold thin enough to be translucent.

And for his ultra-violet filter, to admit the short-wave radiation and nothing else (at about 3200 a.) he chose a thin, translucent film of pure silver. Pettit found that there seemed often to be relation between sun-spot activity and the sun's output of ultra-violet (which varies from time to time by fifty per cent or more), at least during the shorter-period fluctuations. He also found that the ultra-violet varies more or less with, but more widely than, the solar constant of total radiation.

A whole book could be written about the earth and the sun — about all the fascinating and portentous happenings within the sun, and about their terrestrial effects in the way of radio interference, aurora, magnetic storms, and atmospheric storms. But in the limited space here available it may be well to grapple at once (and that briefly) with the baffling enigmas of solar control and worldly weather.

World Weather

The connection between varying solar radiation and other solar disturbances, and such earthly phenomena as magnetic storms, aurorae, and radio, is immediate, direct, and relatively clear. Far less immediate and direct, far more complex and many-sided, is the cause-effect chain between solar disturbances and the weather that we experience on this earth. The connection is there — no doubt of that essential fact. But the complexities, the delaying or speeding up of logical effects, or even their complete reversal, have for years been sorely perplexing to every investigator in this field. Nevertheless, some very definite and noteworthy progress has been made, chiefly in America by Abbot of the Smithsonian Institution and by H. Helm Clayton.

In general, it has been found that temperature, pressure, and rainfall are the meteorological elements most plainly affected by the solar rhythms. In years when sun spots are many (and also, perhaps, in years when the solar constant is high, there is apparently more evaporation from the earth's waters, more cloudiness, more rainfall, and, paradoxically enough (owing apparently to increased cloudiness), *lower* temperature over much of the earth.

Summarizing in the year 1935 many of his ideas about the sun and worldly weather, Doctor Abbot wrote as follows:

... I am painfully aware that the limitations of space and funds, the extensive mass of evidence on which I base conclusions, my own ineptness in its presentation, and the preoccupation of readers with other concerns must all combine to prevent even the most interested of readers from deriving that vivid conviction of the truth and importance of these conclusions which is shared with me by those of my colleagues and friends who are most conversant with the evidence. Nevertheless, I hope I shall not have failed to convince the reader of the following propositions:

- The output of radiation of the sun varies, as proved by simultaneous observations at three stations remote from each other.
- The solar variation, seemingly irregular, really comprises 12. or more regular periodicities, which support successful predictions of solar changes for years in advance.
- The periodicities in solar variation are integral submultiples of 23 years.
- These same and other periodicities which are all integral submultiples of 23 years occur in departures from normal temperatures and precipitations at numerous terrestrial localities. The inference is that solar changes influence weather....

Clayton for many years used solar-radiation measurements in the course of actual weather forecasting in South America and the United States. He found that changes in solar radiation profoundly affect the atmospheric-pressure distribution and wind circulation over the earth's surface, which in turn govern the weather of the world. Day-to-day weather in the United States, for example, is governed by the development and motion of large air masses represented on the weather map by 'highs' or anti-cyclones, the formation of vortices between these air masses represented by 'lows' or cyclones, and the occurrence and movement of the resulting fronts between air masses. But as Clayton pointed out, the highs and lows and fronts that appear on our regional weather maps are originally controlled by larger, semi-permanent highs, and by larger, semi-permanent lows — by 'centers of action.'

When solar radiation increases, Clayton finds that the large polar highs tend to move northward and to increase in intensity, and that the large tropical lows tend to increase their size, intensity, cloudiness, and rainfall. From this alignment we should expect, in temperate latitudes, more variety and more violence of weather during sun-spot maximums. In going over weather records of the past few years, I have noted a tendency in this direction. Great storms were more frequent and severe, it seemed to me, from 1926 to 1930 than they were between 1931 and 1935. If Clayton's findings are thoroughly substantiated, and are explained by adequate theory, they will be of the most fundamental importance in the weather forecasting of the future.

Even if the correlations between sun and weather covered only the last few years, they would be impressive indeed. But records of temperature and rainfall carry the story of the sun-spot cycle and its weather effects far back into the past. First there are the rain- and snow-fall records, extending back to the colonial period in America, that have been carefully kept in southern New England. These records seem to show, from the year 1750 up to the present time, a regular rainfall variation over periods of forty-six years.

This is just double the complete sun-spot cycle; the rainfall cycle apparently reverses itself every twentythree years. Nor is this an isolated instance of forty-six-year periodicity in weather. Abbot goes so far as to say that at many stations weather tends to repeat itself in considerable detail in the course of these forty-six cycles. At any rate, the cyclical rainfall variation is also confirmed in the records of the water level of Lake Huron, Lake Eric, Lake Ontario, and other lakes, some of which have been kept since 1837. And in other conti- nents, also, rainfall has varied directly with sun-spot activity. The cycle is shown by historical records of the flow of the river Nile extending back through the dim past to the year 622. Perhaps the seven fat years followed by seven lean years in Biblical Egypt were linked in some way with a solar period.

Nor does the confirmed record of sun-spot influences on earth stop with recorded history, or even with the unrecorded story of the human race. Doctor A. E. Douglass of Arizona has carried it back to the year 1306 B.C. by measuring the growth-rings of trees. In a moist year a tree puts on a thicker annual ring; in a dry year it has to take in its bark belt, so to speak, and do with a thinner growth-layer. Anyone can find some rough indications of recent sun-spot activity by noting the ring-pattern of a large tree that has been cleanly sawed through. Douglass extended his measurements to the largest and oldest of living trees — the giant sequoias of California — which carry the sun-spot story back to the days when Greece was young and Rome still unborn.

Other investigations have carried the sun-spot record into prehistoric times. During the glacial periods of the Pleistocene Age, hundreds of thousands of years back at the very beginnings of the human race, the glacial ice partially melted in summer; and this melting, together with heavy rainfall, formed numerous lakes. Sediment deposits at the bottoms of these lakes give an idea of the intensity of melting and rainfall each year; and it appears that both rainfall and melting were greater in sun-spot-maximum years. Records of the same sort are available also from the Eocene Age, when nothing remotely resembling a man had ever been seen on earth, when the giant reptiles were just passing, when mammals, the 'family animals,' were just beginning to appear in rudimentary forms.

And Douglass has carried his tree-growth cycles back to the Eocene Age also by examining the rings of petrified trees. It appears that our great sun was perhaps generally hotter in the Eocene Age, and that its variations were more marked than they are at present. A British scientist, Simpson, has even suggested that the various Ice Ages in the Earth's past were themselves caused by spells of excessive solar radiation.

Long May It Blaze

The regularity of the solar periods and the relative smallness of its overall changes give us every hope that the Sun will continue to light and warm the Earth in accustomed, dependable fashion for untold ages to come. But we cannot take this absolutely for granted. Besides the very remote danger of the sun's becoming a nova, there is another danger not quite so remote. As Sir James Jeans puts it:

... Thus if the sun were to become 0.03 magnitudes fainter, this representing a reduction of only 3 per cent in its luminosity, it would arrive exactly at the edge of the main-sequence, and would proceed to contract precipitately to the white-dwarf state. In so doing, its light and heat would diminish to such an extent that life would be banished from the earth....

But as Jeans points out, the sun is far more likely to skirt this dangerous edge, perhaps for millions of years, than to plunge precipitately over it.

If our sun became a white dwarf, our fate would be no less tragic than if it became a nova. In the one case we should be burned; in the other, frozen. Fierce and ever-increasing cold would come upon the world in a few feebly lit days and a few long, moonless nights; summer would wither into harsh winter, not only in temperate latitudes but in the deep tropics as well. The great oceans would freeze solid, the atmosphere would liquefy, and the resulting ocean of liquid air would finally solidify also to enshroud us in a shell of crystalline white.

But on that frozen earth the remote stars would still shine, even more brightly with no atmosphere to dim them. And out near those other remote suns, in their solar systems if any, the failing light of our sun might pass wholly unnoticed. Even if our sun were to explode, it would be only another nova to those far hypothetical astronomers, and they would have no possible means of sensing the catastrophe until hundreds of years had come and gone.

For the frightful explosion of Nova Aquilae mentioned earlier in this chapter actually occurred, not in 1918 when we saw it, but in 1304, before Europe knew that America was. Thus are the sun and its variations linked inseparably with our well-being in so many diverse and obscure ways that scientists have scarcely begun to understand them. As the Earth spins on its axis hundreds of telescopes project the solar image for the curious and speculative eye, and influences more subtle trace their records for future analysis.

What the great Sun does still lies far beyond our control, but no longer can its doings go unobserved or entirely unpredicted. Daily our solar-planetary train makes its eleven-million-mile run through space with greater regularity than any limited. Inklings of the immediate schedule have filtered into the human mind only within recent years; perhaps the future will reveal something of the destination and the Dispatcher. Meanwhile we worldlings, who sprang originally from the leaping solar fires by virtue of a sidereal accident, are bound to the distant locomotive with bands of immaterial steel. May the course lie clear before it, through the ages may it blaze; now and for finite ever are we the Children of the Sun.

Chapter 12

Atmospheric Electricity

The term atmospheric electricity usually is limited to only those electric phenomena of the air which normally we never see, and of which we can have no knowledge without the aid of special instruments and equipment. In its wider sense, however, the sense in which it is here used, it includes, besides all these, St. Elmo's fire, that is, the natural corona, or ghostly flames common on high mountain peaks; the aurora; and all forms of lightning. Indeed in this case, as in so many others, it was a study of the conspicuous that led to the field of the inconspicuous—a field of fertile soil and abundant harvest.

The most familiar and yet most arresting natural manifestation of electricity is lightning, ranging as it does from the exquisitely beautiful to the extremely terrifying—beautiful when far away on a dark night (of course out in the country, not in a city canyon); terrifying when so close that we barely escape.

Lightning and its noisy product, thunder, manifestly were objects of awe and superstition ages before the utmost reach of record or tradition, but identifying them with manageable laboratory phenomena, about which we know the how and the how much, is a very modern accomplishment and an exceeding great one. Through the ages until but a few decades ago each nation, the world over, regarded lightning as a terrible weapon in the wilful hands of their own god. Thus to the Norsemen it was Thor's mighty hammer with which he struck many a telling blow on the powerful demons; and to the Greeks, the thunderbolts of Zeus, that scared them half to death the time they had the Trojans nearly defeated, thereby prolonging the war—to the glory of Homer and the laurels of Virgil. But hammer or thunder-bolt, lightning still is with us, still alarming with the mighty crash and roar of its thunder, still striking and still killing.

The next most conspicuous electric phenomenon, best seen during long winter nights when the moon is far below the horizon and the stars are brightest, is that curious light of the poleward skies that now is the steady glow of a half spent fire, again the leaping flames of a raging conflagration, and anon a gleaming sword—dire omens to many who rarely see them, but mere commonplaces to the Esquimo in whom, because of familiarity, they arouse as little emotion as, for the same reason, gaudy, artificial auroras, in the forms of letters, signs and symbols, excite in the city dweller of today.

Another, and the only other, natural electric phenomenon we shall consider in this connection, is St. Elmo's fire, the spooky, heatless flame that sometimes on dark and dreary ocean nights burns in many places; the tops of masts and the tips of spars of ships far out at sea—the souls, some say, of drowned sailors come to warn their mates of a pending storm. And it occurs on land as well, as unexpectedly and mysteriously as on the ocean.

These, then, are the three conspicuous manifestations of atmospheric electricity; lightning, the aurora and St. Elmo's fire. All are recorded in ancient literature, and, being natural phenomena, all of course go back to the time when the earth itself was void and without form. But to the wise men of old, to the scholars of the middle ages, and to the savants of every land and clime, down almost to our own day, they were mysteries of mysteries, with no known relation one to another, nor to any thing else; manifestations, many believed, of divine wrath, or warnings of dire punishments soon to come, unless forsooth, they quickly and radically mended their ways. Neither was there, nor could there be, any confident hope of ever learning much about the causes of these remarkable displays by merely observing them. However, in this case, as in so many others, knowledge came indirectly through an understanding slowly acquired of wholly different but unsuspectedly related manifestations that could be produced at will by any one, and played, or experimented, with as long as amusement prompted in the one case or curiosity compelled in the other.

The earliest recorded of these revealing phenomena was the attracting to itself of small pieces of straw, feathers and other light objects by a piece of amber on being rubbed by dry cloth, or even by one's hand when free from moisture. This wholly unsuspected, and for ages altogether mysterious, behaviour of amber was known at least as early as the days of Thales, or fully 600 years B.C., but for more than two thousand years after that date we had no further knowledge of static electricity, save only the fact that many other objects besides lumps of amber possess this same mysterious property—mysterious, because the amber looks the same and feels the same after it is rubbed as it does before, and yet the gentlest "caress" of the right sort by its possessor changes it from a mere worthless pebble to a thing of compelling influence and power over many an object it has but to approach—an influence that we ourselves cannot exert, "concentrate" as we may with all our might and all our mind.

No wonder rubbed amber was mysterious to the ancients; no wonder either that it still was mysterious through all the middle ages; and truth to tell any child today can ask questions about it that no one yet is wise enough to answer. Indeed to those of us, and that's all of us, accustomed from childhood to think of fairies and imps, angels and demons, Aladdin's calling of powerful genii to do his bidding by the simple device of rubbing a particular lamp in his possession, or a ring he wore, seems far less mysterious than this awakening of strange powers in a piece of mere amber in precisely the same way. The former we take on faith, and for such things our faith is large, especially when they occurred long ago in a far-off land; the latter we see with our own eyes and must believe, but what to believe for ages we did not know, nor in any ultimate sense do we yet know.

Here, as so often happens, the mysterious was explained by the more mysterious, the little known by the less known. The amber was supposed to have a conscious, sentient life of some sort by virtue of which it performed curious tricks with invisible hands. However, when it was found that many other substances besides amber have the same sort of life property while others did not, the beautiful animistic explanation of the poet and the cloistered philosopher gradually became secretly suspect on the part of those in whom the "why?" of childhood had not been wholly suppressed, and from these few the infection of doubt spread and grew to ultimate flat denial, and that despite the fact that even those who most vehemently denied had no enticing hypothesis to offer in place of the soothing faith they rejected. Wisely they more and more followed the sage advice of Old Omar: "Take the cash and let the credit go"—garnered many a sheaf of observations, with no chasing after strange gods to explain them. And the facts of experiment and observation grew into a multitude, and the multitude was variously classified, and given names for the convenience of those who had occasion to write about them or talk about them.

Thus when there was reason to speak of bodies being got in the state a piece of amber is put in by rubbing it with dry woolen cloth, and need for brevity required a short expression, the term "electrified" was used, meaning made like unto electrum, or electron (Latin and Greek, respectively, for amber), not as it ordinarily is but, by agreed understanding, its state immediately after a proper rubbing—the state by virtue of which it temporarily exerts a pull on nearby objects, often greater than the gravity pull on the same objects by the entire earth.

And since the pieces of straw, paper and the like, that are drawn to an electrified object adhere to it, at least for a time, much as they might stick to it if it were smeared with tar, it was reasonable to believe, as long was believed, that the surface of an electrified object is covered with a real something to which material objects can and do stick. This something, with lots of properties besides apparent stickiness, was called "electricity," almost as soon as a name for it of some sort was needed; and that is the name used for it today everywhere and by everyone, save alone the man who familiarly and deftly guides it and controls it in terms of thousands of horsepower for all sorts of industrial uses. He boldly cuts the Gordian knot of entangling technical terms at a single stroke by calling it "juice."

Early in the course of play and experiment—to a "Cavendish" the same thing—it was found that many objects, such as glass, resin, ebonite and hundreds of others, are easily elec- trified by rubbing, while numerous others, including all the metals, show no awakening, however vigorous the rubbing, unless held and supported, or suspended, by an object of the first class, whereupon they too respond to rubbing just like amber.

All objects then can be electrified. Those of the first class tend to retain their charges—to insulate them. They therefore are called insulators. And as they will not readily part with their own charges, neither will they act as conveyors of electricity from one object to another. The metals on the other hand, and all things of their class, share their charges with objects of their own kind, and act as perfect conveyors of charges from object to object. Hence, they are called conductors. Out of these play-experiments (they are a lot of fun) it soon was found that two things electrified alike, two bits of ebonite, for instance, rubbed with the same piece of fur, though equally picking up all sorts of light bodies, nevertheless repel each other. But they do not repel all electrified objects.

Some they more strongly attract than they do things that are not electrified. Hence we conclude that there are two kinds of electricity; that quantities of like kind repel each other and that quantities of unlike kind attract each other. And here is another bit of informative fun: Wrap a piece of cat skin (cheap and effective, that's why it is used in physical laboratories) around a rod of ebonite, fur-side inside, fasten the bundle to an insulating support and twirl the rod. From what we had found before by rubbing, ebonite with fur, the rod at least should now be electrified. However, the bundle as a whole gives no sign whatever of any such condition. Perhaps we pull the rod out to see what the trouble may be, when presto, both skin and rod are fully charged. So we put the rod back again and just as presto the signs of electrification are all gone.

We pull it out—charged so on as often as we like. A little more play and we find that the charge on the rod is of the opposite kind from that on the fur; and since before the rod is removed from the fur the two together show no signs of being electrified, we conclude that not only is the electricity on the rod different in kind from that on the fur, but exactly equal to it in quantity. Innumerable other experi- ments, many of great delicacy, have been made to determine whether one kind of electricity ever is obtained without at the same time getting an equal quantity of the opposite kind, and always the result was the same—never one without the other, simultaneously and in equal amount.

Thus, the never-grownup play-boy, bubbling over with curiosity to try new tricks with new toys, learned many things from his closest and dearest playmate, Nature, and from her he alone will continue to learn, he the one and only Prometheus.

Anyway, among innumerable other facts he (in this case Sir J. J. Thomson) found that, as long suspected, this electricity is as granular as the sands on the seashore, but incomparably more uniform, in fact entirely indistinguishable one grain from another when of the same kind. The granule of the kind commonly known as negative electricity he named electron, and the equivalent granule of positive electricity proton. And in ways delightfully simple and highly entertaining, he found out just how much electricity there is in each of these grains, each electron or proton, and it is mighty little in terms of the units of the electricity we use for light, heat and power. Thus, it would be something of a job to count the number of electrons equivalent to the electricity used in lighting for only a minute or two an ordinary 60 watt lamp, which takes a current of roughly half an ampere and is just sufficient to read by when close to it—an impossible job, for it would take ten million people, each counting continuously at the rate of 100 per minute, 100 years to count enough electrons to run this lamp just one second of time!

Soon after the discovery of the electron, it was found that the atmosphere is full of electrons and protons, or electric dust if we wish to think of it that way, a state that explains many phenomena that otherwise would have to be accepted as just so many disconnected facts, so many scattered beads, as it were, and not a string of beads held together by a common thread.

To follow this subject farther it is desirable first to get a clear understanding of the meanings of a few technical terms that are so convenient as to be virtually necessary, terms in everyday use in connection with domestic and industrial electricity, the electricity used in lighting houses and running fans. This gentle sort of electricity (there is a violent sort that will be discussed presently) was played with and even trained to do many useful things long before the discovery of the electron.

One astonishing way of obtaining it (they are all amazing) is by forcing a wire crosswise into, or out of, the space between the poles of a magnet. During, but only during, the motion of this wire its ends, if it is discontinuous like a rod, are electrified oppositely, in one sense while the wire is being pushed into the space between the magnetic poles and in the reverse sense while it is being pulled out again; and when the ends of this wire are connected with an outside wire loop, an alternating current flows through the conducting circuit as any section of it is forced back and forth between the poles of the magnet, or even across the face of just one pole.

Exactly why these things happen is too long a story to tell here and not a simple one either. However, all dynamos, toy type to complex and powerful, are based on this one simple phenomenon. Another amazing electrical discovery, made before the one just described, and started by the spasmodic kick of a dead frog's hind legs, was that when certain pairs of substances are immersed, or partially immersed, in particular fluids and then connected by contact, or even with a wire quite outside the fluid, chemical activity on one of the immersed substances is immediately evident; that the wire becomes warmer; that the pointing of a nearby magnetic needle is changed, and that the longer this wire (of the same material and same size) the feebler the effects.

These and many other facts of observation and experiment soon led to the notion that electricity is a something analogous to a fluid, in that it call flow along the substance of a wire like water through a pipe; that the wire resists to a greater or less extent the flow of the electricity as a tube offers more or less resistance to the flow of water through it; that in each case there is a current; that the strength of this current increases with the push that drives it and decreases with the increase of the resistance it encounters.

The electric push we call voltage, and measure it in terms of an appropriate unit. The opposition to the flow of the current we call electric resistance. The volume, or strength, of the current we call amperage, and the unit strength an ampere. The unit quantity of electricity is called a coulomb. Of course, there also are other names for this and that electrical phenomenon and its measurement, but the above will do for the present; additional ones will be explained incidentally when needed.

But back to our muttons, atmospheric electricity. In performing tricks with electric toys one often has occasion to set a charge on a pedestal and have it stay there without change until needed, or to bottle it up like the chemist does his reagents. But this much-sought-for result has never yet been attained to perfection; and fortunately not, for the failures taught us far more than success, if that had been possible, would have revealed. At first one thought to imprison a charge by supporting the electrified object, a metal sphere, say, or metal shell (just as good, for the charge is all on the outside anyway), by an insulator, such as a glass rod. Sometimes this device was rather effective, and sometimes it just didn't work at all. It was effective only in dry air and wholly worthless in highly humid air. A sulphur support is less sensitive to humidity, but all fail in hot, muggy weather.

This fortunate failure started a bit of detective work to locate what in Boston, they say, is called "the Senegambian in the cellulose heap." Pretty soon he was found, at any rate one was found, and proved to be humidity on the surface of the "insulating" support, by virtue of which this surface becomes a good enough conductor to permit the charge to flow away immediately. This humidity, however, does not appear in the form of dew, such as often gathers on the ice pitcher; rather it is present in a quasi solution of the superficial layer of the support.

The more hygroscopic the substance of this support, that is, the greater its tendency to absorb, or unite with and retain moisture, the worse it is as an insulator in humid air. Clearly then, or so it naturally was inferred, an electrified conductor will retain its charge indefinitely if it is mounted on a thoroughly dry and perfect insulator. Wrong again! True, the better the insulating property of the support, and the drier its surface, the less rapid the loss of charge, but still there remains a steady loss of charge that soon—a few hours, perhaps, to a few days at the very best-completely exhausts it. How to account for this loss of charge from an electrified body in the atmosphere appeared to be so easy and evident as barely to deserve mention; just a case of dust, and perhaps some of the adjacent air itself, getting charged by contact with the body in question, as long as it had any charge to share, and then, because the charges are alike, being driven away by their mutual replusion—a process of attrition through the dissipation of an "infinite" number of infinitesimal bits.

A beautiful theory this, simple, satisfying, supported by experiment, generally accepted and universally taught; yet all wrong, or nearly all. A little of the charge normally must be lost this way, but a portion so small as commonly to be quite negligible. Evidently, if this were the way an insulated object loses its charge, that loss would be more rapid in smoky, or dust-filled, air than in clear air, whereas it is slower.

Again, the rate of loss should be the same for the two kinds of electric charges, but they are not equal. And so, for these and other good and sufficient reasons, the above beautiful theory was cast aside, but without regrets or confusion as a far better already was at hand, one that not only explained all it was meant to explain, but also led to new discoveries. At first it was just that air is a conductor of electricity; and now that it is it conductor by virtue of being filled with a vast number of minute electrical charges of both kinds, positive and negative, as we call them, or + and -, in the convenient shorthand of symbolism. An insulted charge object, then, attracts to itself the surrounding opposite charges, or oppositely charged particles, and thereby cancels these many minute charges with an equal amount of its own supply until that is all exhausted.

But this theory raised a lot of questions: Are these little charges disembodied ghosts, or are they attached to ordinary matter of some sort? Are they all of equal quantity, or of different quantities? If attached to matter what is the state or condition of that matter? How many such charges of each kind normally are present per cubic inch, say, in the lower atmosphere?

How does this number vary with season, altitude, place and state of the weather? Does the conductivity of the atmosphere vary? If so why? Where do these charges come from and what becomes of them? Such are some, but by no means all, of the questions that have been asked in relation to atmospheric electricity—and answered, in the main. Some of the answers were gotten by putting two and two together in this way: It long had been known that a little sphere falling freely through still air quickly acquires a uniform velocity whose value is determined in a known way by the size of the sphere, its weight and the density of the air through which it is falling.

It also was known of old that an electrified object between two oppositely charged parallel plates experiences a force normal to the plates, directly proportional to its own charge and to the difference in voltage between the plates. Very well then: Note the rate of fall of a spray-droplet of a given non-volatile liquid between two uncharged, parallel, flat, metallic plates in perfectly quiet air of known density and temperature, and thereby obtain the size and weight of the droplet. Then, while the droplet is in the air between the plates charge them to a known difference in voltage, and note the then rate of fall of the particle. If its speed remains the same after the plates are charged as it was before, there obviously is no electricity on the droplet; but if this speed does change, the droplet then just as definitely is charged, and the amount of this charge is immediately computable from the data thus obtained. In this straightforward way it was found that such droplets sometimes are neutral, at other times negatively charged and at still others positively charged. Furthermore, the quantity of the charge varies, but always is equal to, or some whole-number multiple of, the minimum ever obtained. This minimum quantity is the electric atom, the value of the electron charge, and of the proton charge, for they are equal to each other.

Delightfully straightforward experiments these, but, when carried out with all the necessary refinements, worthy of the Nobel Prize they won for Millikan.

The electric grains, or atoms, therefore, are all of the same magnitude but of two kinds. However, they may gather in multiples on a common object, and the presence of either sort completely nullifies an equal amount of the other sort.

Each kind normally is associated with ordinary matter, though it can be forced to a separate existence. However, it even then retains inertia and therefore one of the basic properties, if not *the* basic property, of matter. Is then every bit of matter, mote to mountain, just an aggregate of positive and negative granules of electricity, or is electricity matter in a special state? We don't know. Indeed we who ask these questions generally have very hazy ideas as to what we mean by them anyway. It will be safer, therefore, to avoid the bogs and quagmires of this philosophical detour and keep to the hard-surface highway of things tested and true.

We know that of electrons, or protons, it would take more than twice (actually about 2.1 times) a thousand, thousand, thousand, all at a single point, to push another equal charge at a point one centimeter distant (one inch = 2.54 centimeters) with the force of one dyne, or the one 980th part of the weight of a gram, and it takes 28.35 grams to make an ounce. And this quantity of electricity that pushes an equal charge at the distance of one centimeter with the force of one dyne, is called the unit quantity of static electricity. Clearly, then, an electrified body such as a metal sphere, suspended in the free air by a nonconducting fibre, or supported on an insulating stand, would slowly lose its charge, as such bodies do, even if the insulation were perfect, if there were freely moving charges of the opposite kind in the adjacent atmosphere.

To determine whether or not the air does contain such free charges, one may set up a metallic pipe connected to the ground with a wire, or other conductor, so it cannot be charged, and place along its axis, but not extending its whole length, an insulated metal rod, charged and connected to an instrument (electroscope) that shows the intensity (voltage) of this charge. Then draw air at a known slow rate through the pipe over the charged axial rod and note the rate at which it loses its charge.

By a little experimenting one soon finds in this way the loss of charge per cubic centimeter, say, of the passing air, and from that the number of electron charges of opposite sign from that of the charge on the rod per unit volume in the free air. In this general way, with various modifications and refinements of apparatus, it soon was found that after the air is seemingly cleared of all its electricity there still remain usually several to many times as much that yields far more slowly to an electrical attraction on it. The more easily removed portion is attributed to light or mobile ions—charged molecules, presumably—and the more resistant part to relatively massive ions, such as charged condensation nuclei and motes of whatever kind.

The relative slowness with which the two classes of ions, light and heavy, come out under the same electric pull appears to be owing to their respective mass inertias, the heavier being correspondingly hard to speed up. The numerical density of the positive light ions is about one-fifth greater than that of the negative, and averages around 800 per cubic centimeter (roughly 12,000 per cubic inch) in the lower atmosphere over both continents and oceans. The positive and negative heavy ions are about equally numerous, but while relatively scarce over seas, roughly 200 per cubic centimeter, they are far more numerous over land where they range from about 1,000 per cubic centimeter to nearly a hundred times that number, or, say, 1,500,000 per cubic inch. The number per unit volume of heavy ions varies with the season, latitude, time of day, state of the weather, direction and distance from cities, and a variety of other circumstances, all of which contribute to the one controlling factor, namely, the "dustiness" of the air, for it is charged "dust" particles (including fog or cloud droplets) that are the heavy ions.

Because the atmosphere is filled with thousands of both positively and negatively charged molecules and motes per cubic inch, it therefore is an electrical conductor. Evidently, too, its conductivity increases with the numerical density of the ions present and with their mobility. It would seem also that this conductivity must decrease, in general, and in general it does, with the dustiness and fogginess of the air, since the lighter and therefore more mobile ions are nearly all captured by the relatively massive and sluggish dust particles and fog droplets. This explains, too, why the electrical conductivity of the atmosphere varies with location, season, time of day and state of the weather, for it is with these conditions that dust and fog vary. It explains also why the conductivity increases with height, as the higher the less the dust, and the finer the particles.

Since, now, the atmosphere is full of minute electric charges of both kinds, not only are there electric forces between every one of the myriads of millions of pairs of such charges, but also pulls between the charges and the earth beneath. Whenever, then, as almost always on clear days, there is more positive electricity in the air per cubic inch than negative the air is said to have a positive space charge, that is, more positive electricity than negative dispersed through the air alone (to some extent in the case of electrons) or on motes, molecules and atoms.

And since, for every charge of one sort of electricity there always is an equal charge of the other sort as nearby as possible, there must be, in this ease, a negative charge on the earth beneath equal to all the excess positive charge above, and as the earth is a conductor this negative charge must be on its surface. Clearly, therefore, an object in the atmosphere will be pushed upward if electrified like the surface of the earth, and pulled downward if electrified in the opposite sense. In either ease there would be an electric force on an electrified object anywhere in the free air.

Sometime, we have occasion to measure this force and often to talk about it, and when we do either it is convenient to use agreedupon units and technical terms. Thus, we speak of the electric potential at a given point above the surface of the earth, and mean thereby the work that would have to be done against the electric forces in taking a unit quantity of electricity of the opposite sort from that on the earth beneath up from its surface to the point in question; and the difference in potential between two points to be the similar work required to take such unit of electricity from the lower to the higher point, commonly directly above, but not necessarily so.

Obviously, when the surface of the earth is electrified one way and the body of the air above oppositely, then with increase in height the more nearly do the quantities beneath, the surface charge and the volume charge, become equal to each other, and the feebler the force on a given charge. Hence, there must be a tendency for electricity to flow from bottom to top, or top to bottom, of any object of appreciable height. If it is a conductor, iron, say, and several hundred feet high, there occasionally will be a current along it that leaves (or enters) the highest point as a coronal, or brush, discharge-St. Elmo's fire.

Even when the height of the conductor is only a few feet, it is easy with suitable instruments to measure the electrical push along it, or difference in voltage between its upper and lower ends. In clear weather this push in the lower atmosphere amounts to about 100 volts per meter difference in height. Therefore, between one's head and feet on the golf links, or elsewhere in the open, there commonly is an electrical push, or difference in potential, about double that which forces the current through our electric lamps. But the voltage is one thing and the quantity of electricity behind it quite another, and in this case the quantity available is far too small even to be felt.

As stated above, the electric force on a charged object necessarily decreases with increase of height, and therefore, in the same proportion, so does the rate of change of electrical potential.

It is obvious also that if the rate of supply of electricity to the free air is roughly constant, as it seems to be, except in thunderstorms, then the better conductor this air is the less accumulation there can be of electricity in it and on the surface of the earth and the less rapid the change of potential with increase of height. That is why this potential gradient, as it is called, commonly is greater in a fog, or when the air is smoky or dusty—when the electric carriers are relatively massive—than it is in clear weather when the ions average much smaller and lighter. This effect of the "dustiness" of the air on the potential gradient explains in part why it varies through the day, for that is what the dustiness of the lower air does, and why over continents, especially during the summer in mid latitudes, it commonly has two maximum values, one about 8 o'clock in the forenoon and another at 6 to 8 o'clock in the evening.

This double diurnal change in the potential gradient is only a shallow, continental effect. It does not occur at heights above 1,500 feet anywhere, nor at any level over the oceans, nor in polar regions, places where there is neither smoke nor dust, nor much convection. In these places, however, as everywhere else, there is a diurnal range in the potential gradient of, roughly, from 20% above its mean value to 20% below. This change appears to be wholly independent of dust, fog, temperature or anything whatever of a local nature, for it is of about the same magnitude the world over, and everywhere is concurrent, with the minimum value at about 4 o'clock A.M., Greenwich time, and the maximum at about 8 o'clock P.M., also Greenwich time. This coincides with the diurnal variation in the world's number of thunderstorms. The obvious connection here is the fact that lightning alters the magnitude of the charge on the surface of the earth, which change instantly spreads everywhere, as the earth is a conductor.

ELECTRICAL STATE OF THE HIGH ATMOSPHERE

It is known now from the phenomena of wireless transmission, that the very high atmosphere is greatly ionized, and that the amount and distribution of this ionization widely change between day and night. The telling of what these changes are, so far as known, explaining how they are measured, and speculating on their causes, make fascinating reading for the specialists who are engaged in, or at least understand, that kind of investigation. But the story is too long to attempt here, and besides would be out of date before it could reach the reader, so rapidly is it being developed.

It seems to be pretty well established, however, from the phenomena associated with wireless transmission, that the upper atmosphere is much more strongly ionized during the daytime than it is at night, and more strongly when sunspots are numerous than when they are few. Furthermore, this ionization is largely concentrated at certain levels, especially at about 73 miles and 160 miles, respectively, above the earth. This upper layer, however, appears to be quite variable in height and often even to divide into two layers of considerably different heights. Such are some of the facts of observation, and there are many others, but how they are related, one to another, and exactly how they are caused we do not yet know, and may never know fully, though day by day we shall surely come nearer to that complete understanding.

We do not know, for instance how, nor from what source, the earth gets her continual vast charge of negative electricity despite a steady out-current of 1,000 amperes. There is good reason to believe that this state of the earth's surface is largely owing in some way to the action of thunderstorms; partly to emanations from radioactive elements; and partly, also to cosmic rays from "Lord-knows where"; and partly, perhaps, to many other sources. We know, furthermore, that the ever-flowing air-earth electric current, if undiminished, would completely discharge the entire earth in 7.5 minutes. Whence, then, comes this current, ceaseless as the fall of a cataract that never runs dry? Quien sabe?

SAINT ELMO'S FIRE

Nothing is more awesome and soulenthralling than the unbelievable realities of a dark and stormy night on a tropical sea. Beneath, and all around, a vast and glowing fire that consumes not nor is consumed; and above, on every higher point, a cold, ghostly flame of sulphurous odor. The first of these mysteries, occurring, as it does, beneath the water surface, is not in any sense a manifestation of atmospheric electricity, and therefore, no discussion of it will be offered here. True it is likely to be exceedingly active whenever the tipping flames occur, but not owing to any causal relation between them, for they are of wholly independent origins—it is just that a storm intensifies the one and may induce the other, and that the two together cast a deeper spell than could either alone of awesome mystery.

The flames, however, that blaze without burning on the topmost points, or jump from place to place like torches thrown by invisible hands are wholly electrical—just one of the beautiful things the physicist loves to play with in his toy house, the laboratory. There he sets a suitable electric machine running in the pitch dark, and presto it is all ablaze with bushy flames, "brush discharges," he calls them, from a hundred places. Tufts of frazzled lightning is what they really are, even if that term does seem a trifle disrespectful of Jove's mighty thunderbolts. The same phenomenon appears also along certain power lines that carry deadly high-potential currents. In this case the glow caused by the loss of electricity from the wire into the air is called a corona, and the loss itself a coronal discharge.

But "brush discharge" and "coronal discharge" are recent technical terms used by the physicist and the engineer in speaking of the loss into the atmosphere of electricity from objects charged to such high potential that the resistance of the air breaks down under the strain. Exactly this same phenomenon, caused in the same way—a brush discharge due to great electric stress (high voltage) occurs from time to time, and especially during the passage of a thunderstorm, on the high points of ships at sea, and on land from tall towers and particularly from the tops of isolated peaks.

THE AURORA

To one who had never seen nor heard of such a thing a glowing red midnight sky might portend the crack of doom; and even more appalling would be a sky hung with luminous draperies, fold upon fold; and terrifying in the extreme a forest of gigantic flaming swords, furiously slashing across half the heavens. And yet to all this, and more, the Esquimo pays not the slightest heed. To him the aurora in its many forms and various colors is a familiar sight.

We are certain he has no idea what it is, for he explains it in much the same way that we did, not so long ago, but he does know that it is neither harmful itself nor the harbinger of evil, and so to that which in many another would inspire awe, at least, and perhaps even terror, he gives not the remotest thought. Familiarity begets indifference. As just implied, the aurora is a night-time glow in the sky that more or less resembles the light produced by an electric discharge through a rarefied gas. No doubt it occurs in the daytime quite as often, on the average, and as brilliantly as at night, but it is not then visible because the brightness of the day-sky is so much greater than that of the aurora that the glow of the latter is completely lost in the glare of the former. Neither can it be seen when the sky is completely overcast, because its closest approach to the surface of the earth still is far beyond the loftiest cloud.

In respect to region of frequent occurrence there are two classes of auroras, called, respectively, northern lights, northern aurora, or aurora borealis; and southern lights, southern aurora, or aurora australis, according as reference is had to its appearance in the northern or southern hemisphere. When, however, reference is to the shape of the luminous region, as seen against the heavens, auroras may be divided into arcs, bands, rays, curtains, coronas, patches, diffuse glows, and as many more as one's fancy can suggest and his vocabulary supply—mostly useless. In the northern hemisphere the aurora is most frequent at about latitude 60° over North America and the Atlantic Ocean, and near 70° in Siberia. In short, the locus of its most frequent appearance in the northern hemisphere is, roughly, a great elliptical ring around the north magnetic pole, but much nearer to it on the American side than the Asiatic. Away from this ring the occurrence of the aurora is rapidly rarer with decrease of latitude. In the region of the Great Lakes, for instance, as in Scandinavia, it is quite common; unusual in the Gulf States and along the Mediterranean; and rarely seen within the Tropics.

As already stated the aurora can not be seen against the glare of a sun-lit sky. Hence, to the eve it is strictly a night phenomenon, like the Milky Way, and for the same reason it is more frequently observed during the winter, when the nights are long, than in the summer time when they are short. In addition to these diurnal and annual periods, of obvious cause, in the appearance of the aurora, it also has an eleven-year period, coincident with the sunspot period in the sense that when the spots are most numerous the auroras are most frequent, and rarest when the spots are fewest. It, therefore, parallels also, at least roughly, the occurrence of magnetic storms. It has no relation to season, beyond that incident to the length of the night; none to the phases of the moon except that faint auroras cannot be seen in bright moonlight; nor any relation to the weather, except the obvious one that it never is seen when the sky is overcast. Evidently not, for clouds never rise much above 6 miles in auroral regions, while ten times that generally is the polar lights' nearest, and usual, approach to the earth—nearest approach, not its outermost reach, for that is anything up to 600 miles, or more. Surprisingly, too, it is visible, and may be photographed, much farther out when its topmost portion is in sunshine than when it is in the shadow of the earth, of course when the sun is way below the horizon and the lower atmosphere therefore free from obscuring glare.

As already stated, the aurora is of various colors: silverwhite; pink; red—pale to deep and dense; green, of the mother- of-pearl or nacreous appearance; and yellowish. Often the same streak is red in its lower portion and greenish in its upper reaches. In shape, colour and motion the polar lights play varied and fascinating roles, all incident to electric discharges in the very high, rare and exceedingly ionized atmosphere. Most of this light long ago was traced to nitrogen, and recently, after many defeats, the rest of it to oxygen.

The more brilliant auroras commonly are accompanied by disturbances in the magnetism of the earth; also there are at these times marked electric currents in the earth, strong enough occasionally to interfere seriously with telegraphing. These facts make it all but certain that the auroral lights are caused by electric currents in the high atmosphere. This conclusion is further supported by the fact that the streamers of light follow the lines of the earth's magnetic force, just as an electric current would; and by the clinching fact that practically the only way to make oxygen and nitrogen give out light, the way they do at great heights, is by passing an electric current through them. That these luminous streamers parallel the earth's lines of magnetic force is shown by the fact that they all point to some place in the direction of the magnetic pole.

When there are several of them at the same time they seem to fan out from a common hub, or pivot, like streaks from the sun, only in this case the hub turns out to be the magnetic pole, and the fanning more apparent than real—a perspective effect like the seeming drawing closer together with increase of distance, of the parallel rails of a straight stretch of railway. This analogy is close, for the streaks are as straight as the rails, and the portions visible from any given place nearly parallel since their common hub usually is thousands of miles distant.

It also is known, from many years of observations, that auroral displays commonly are brightest and most extensive when a great sunspot squarely faces the earth, a fact that forces us to assume that the electric current in the extremely rarefied outer air that produces the aurora comes directly from the sun. And we believe, too, that this current is a stream of electrons—atomic negative electricity—or, occasion- ally, perhaps, positively charged atoms; for only such streams cause oxygen and nitrogen to glow as they do in the aurora.

How the heights of the auroras are measured deserves special and separate mention for its very cleverness and simplicity. Many auroras have clearly distinguishable, temporary markings or splotches, and often a fairly well defined and more enduring base. Evidently, therefore, it would be possible for observers at two stations, say 20 miles apart, in a known direction from each other, to record simultaneously the directions, horizontal and vertical, from their respective positions, of the same spot in an auroral streamer, and then with a little calculation of the easier sort compute its exact height above the earth.

In practice, however, the necessary observations consist of simultaneously photographing the aurora at the two stations against a background of stars (always visible through an aurora), and noting the exact time the exposures were made. Clearly the pictures of this same spot in the aurora will be at one place among the star images on the negative obtained at one of the stations, and at a different place on the negative made at the other station. Then knowing these apparent positions of the spot in question among the stars, the exact time of the exposures, and the precise geographic locations at which they were made, one has only to turn to a suitable astronomical table to get the direction of the given spot from each place of observation, and quite as accurately as could be obtained with a theodolite (telescope provided with horizontal and vertical graduated circles). Finally, from these directions, and the direction from one station to the other, together with their distance apart, it is a simple and easy matter to compute the height of the given spot in the aurora.

This photographic method of determining auroral heights is longer than that by direct observations, but it obviates all the difficulty of obtaining simultaneous measurements, and all uncertainty as to whether the records of both stations apply to exactly the same point. It also enables one to study every distinguishable detail, and that, too, calmly, carefully and at any subsequent time. In short, though more tedious, it is much the better of the two methods.

LIGHTNING

The enchanting tracery of lightning over a distant summer-night's cloud; its blinding brilliance when alarmingly near; and the rolling rumble that goes with it, all, from ancient days, are recorded in history, related in legend and told in story. Its splendor, mystery and danger ever evoked admiration, reverence and awe. It may never itself have been an object of worship, as the sun has been and is, but momentarily, at least, it many a time compelled pious and prayerful thoughts—only that and nothing more.

It could not be gotten hold of like sticks and stones and thereby known as they were known, and so age followed age with no attempt whatever to account for lightning as a natural phenomenon. Nor, indeed, until comparatively recent times was there any possible way by which such an attempt could rationally be made. Sculpture, architecture and poetry had attained levels never yet surpassed, ages before any one had the slightest understanding of lightning; and all the world's great religions long had been preached and practiced, so much earlier, and more effectively, did we contemplate than investigate. The injunction of the ancient poet and philosopher was, "know thyself"; that of the modern scientist is, "know thou Nature." Both are productive of infinite good, but it was the latter, and it alone, that changed lightning for us from a terrifying, wilful act of an offended god to a normal and understandable natural phenomenon. For the former there were reasons aplenty; for the latter not a scintilla of evidence.

The first recorded hints from that great teacher, Nature, as to what lightning is were made, as her hints always are made, to those who *saw* what they saw—who comprehended the possible significance of what happened before their eyes. In this case Hawksbee published in 1705, and Wall in 1708, accounts of their experiments with frictional electricity, and how the phenomena thus obtained, though tiny and feeble, strangely resembled thunder and lightning.

From such experiments as these, made by many persons, it was surmised that thunder clouds are charged with electricity and that lightning is a manifestation of their discharge—surmised, but nothing more for as yet no experiment had been made that had any bearing on this problem, nor any even devised that would at all test the matter. In 1749, however, Franklin, proposed a simple means "to determine," as he said, "the question, whether the clouds that contain lightning are electrified or not," but had no opportunity at that time, nor for some years thereafter, to convert his paper plans into physical equipment and subject it to the test of an actual experiment. But the plan itself had become well known, and on May 10, 1752, the Frenchman, Dalibard, proved by it that the clouds in which lightning occurs are electrified, and during the same summer Franklin confirmed this conclusion in a more convincing way —with his famous kite experiment.

There the matter rested, or practically just there, for decade after decade, until the scientist's curiosity induced, the industrialist's needs required, and new devices made possible, a fuller knowledge of lightning in its every phase, from genesis to dissipation. The search is on, keener than ever before, and more varied for there is no such thing as satisfying the investigator's curiosity—answer him one question and he asks a dozen other. Thus we learn.

If now we are going to talk about lightning, and we are, it will be well to follow the logical order in the famous receipt for cooking a rabbit pie, and first provide a cloud with the necessary electric charge. The fact that lightning occurs in a certain type of cloud and in no other, nor in that except when in an active state of precipitation, is so conspicuous that our veriest ancestral savage must have known it. But he did not know, and did not care, nor until recently did we know, however impatient our curiosity, how any cloud became electrically charged, nor what precipitation had to do with it.

One of the first leads, perhaps the first important direct lead, to the understanding of this phenomenon, as it now is more commonly explained, was found in 1892 by the German physicist, Philipp Lenard, when he observed that pure water, when splashed against a solid object, becomes positively charged and the near-by air negatively charged. This lead was followed to an open highway in 1908 by the English meteorologist, George C. Simpson, then in India, when he showed that the rupturing of drops of distilled water (ordinary tap water will not do, to which hangs a tale all its own) by a blast of air produces the same Lenard effect; the finer spray particles and the adjacent air being negatively charged, and the remaining larger parts of the drops positively charged. By George, be said—at least that is what any enthusiastic investigator would have said—that is how the thunderstorm cloud gets electrified. Rain water is distilled water, just the sort needed; check one.

Then, too, the uprush of air in a thunder cloud (cumulonimbus), as evidenced by the occurrence of hailstones, and testified by the aviator, is strong enough to keep raindrops suspended until they grow so big, about one fifth of an inch in diameter, that they are shaken to pieces by the wind that sweeps past them and carries the spray, at least, to still greater heights, an action that in the laboratory produces electrical separation; check two, and a long way, it would seem, on the road to a complete demonstration. But admitting, as we must, for the evidence is conclusive, that the air in a thunder cloud does rise with surprising velocity we naturally want to know why it does so, and at what level it will be breaking up raindrops at the maximum rate and thereby most rapidly building up the amount of electrical separation, or magnitudes of the electric as we commonly say, positive and negative. Well, it is just the gravity phenomenon of the heavier, or denser, pushing the lighter up, as water pushes up a cork that is let go beneath the surface.

Commonly this difference in the densities of adjacent masses of air, essential to convection, is effected in one or the other of two quite different ways. One is by local surface heating by which the air at the warmer place is brought to a higher temperature, and therefore is more expanded and lighter, volume for volume, than that round about. The other is by the importation over the surface of air colder than that surface. Increase in the amount of water vapour in the air also decreases its density, but generally this is of relatively little importance and therefore will not here be further considered.

In any case the lighter, or more exactly, the warmer air is pushed up by the colder, because it is the denser of the two. As this warmer air is forced up to higher levels it comes under less and less pressure to the extent of the weight of the air left below; and being under less pressure it expands, of course against whatever pressure may still be on it.

But this expansion against pressure means work, and work at the expense of the energy, or ability to do work, of the expanding air, that is, at the expense of its heat. It therefore cools with ascent, and often before it has gone a mile high its temperature has fallen to the dew point and condensation has begun. This condensation, in turn, sets free the heat that was used up in the first place in evaporating so much water as is now changed back from vapour to liquid, a surprising amount, approximately enough for every pound of cloud formed to heat six pounds of water from the freezing point to the boiling point. Yet, paradoxical as it may seem, this tremendous amount of heat does not increase the temperature of the air to which it is given, but it does prevent that air from cooling as rapidly with increase of height as it otherwise would, and thereby keeps the interior of the cloud all the time warmer than the air on the outside at the same level.

In this way a vigorous chimney-like up-draft is created within a cumulus cloud, maintained so long as there is an abundant inflow at the base of adequately humid air, and carried up to that height at which, incident to the resulting low temperature, but little vapour is left, too little, indeed, to furnish enough heat, through condensation, to push the convection to greater heights. The level at which the up-draft is strongest doubtless varies greatly from cloud to cloud, but from the fact that every foot of ascent develops an urge (though a progressively decreasing one) to further ascent, and also from the direct testimony of aviators, it seems that this level of swiftest up-rush normally is well above the cloud's midheight. The vigorous uprush of air in a thunderstorm cloud, therefore, is fully accounted for by the great amount of heat freed by, or resulting from, the abundant local condensation. Furthermore, we know too that the velocity of this ascent through much of the cloud's height is quite sufficient, 30 feet per second, or more, to carry along all water drops of smaller diameter than about one fifth of an inch.

Any larger drops that may form in this vigorously ascending air are blown to bits, just as such drops are in falling through still air on attaining a velocity of about 26 or 27 feet per second. The fragments broken off from the larger drops by the ascending current, are, of course, carried on to still greater heights. Hence the tendency is to accumulate a great amount of water at that level in the cloud where the ascending air just supports the largest drops that can remain whole. That is, the level at which the velocity of ascent is 25 to 30 feet per second.

This level appears usually to be above the midheight of the cloud, and since it necessarily is the place where electrical separation mainly occurs, and where the positive charge is chiefly concentrated, it follows that ordinarily most of the positive electricity in a thunderstorm cloud is in its upper portion. The negative electricity carried off on the spray from the ruptured drops canopies the top of the cumulus and, owing to the descent of the air all around it, drapes its sides. Hence a cloud charged in this way may be mainly positive in its lower portion, and negative in its upper portion, as previously was thought to be its usual state; or the other way around, positive above and negative below, as observation shows it generally is and as appears to be an entirely possible, if not, indeed, much the more likely result of electrical separation through the rupture of raindrops incident to vigorous convection.

As just implied there still is some uncertainty as to the exact process by which the towering thunderstorm cloud becomes electrified, but it is certain that such clouds do become tremendously charged, one portion with positive electricity and another with negative electricity. The two charges may be separated any distance, just as may artificial charges may far or near they never let up their pull on each other, partly through the nonconducting air directly, and indirectly, in part, through the conducting earth beneath. But close together with nearly all their mutual pull direct, or so far apart that the pull of each is essentially between itself and the charge it induces on the surface of the earth below, neither can be increased indefinitely any more than can a like charge on an object in the laboratory.

As the charge becomes greater and greater the electrical resistance finally breaks down—the crowded electric atoms burst their bounds much as gas molecules burst the walls of an over-filled balloon. As the charge builds up, or becomes more and more concentrated, any free electrons present (and there always are some) are moved with greater and greater speed by the force the increasing charge exerts upon them and in the general direction of the nearest large charge of the opposite sort. At first the impact of these electrons on the air particles (molecules) just bumps them a little but leaves them otherwise unaffected.

As the impelling charge grows greater, however, some of the electrons move with such speed that they no longer just rebound from the molecules they encounter, but disrupt them, at least to the extent of driving off one or more of their normal number of electrons. In this way the conductivity of the air into which the electrons are driven, or "shot," rapidly becomes greater and greater, the free electrons, incident to the increased bombardment, still more numerous and faster, and so on and on, every increase of either condition intensifying that of the other, until, in less time than it takes to bat an eve the selfboosting climaxes in a spasmodic rush of the whole charge, or much of it, at least, along a sinuous and often multi-branched narrow path which all but instantaneously becomes intensely luminous, and which we, therefore, call lightning, meaning, if we mean correctly, a flash of light induced by a natural electric discharge through the air.

The flash of light, as stated, is not coincident with the passage of the electricity, but is a consequence thereof. The discharge itself consists of extremely fast moving electric, or electrified, particles, electrons mainly, along a very crooked tube-like path. Along this tube, then, there are countless myriads of both free electrons and roving positive atoms—positive because partially stripped of their normal supply of electrons, with which they are neutral.

Instantly this separation of positive and negative charges is effected, they start recombining, each electric atom with its original partner, forming the same old neutral atom, or, more likely, with a different partner, and forming thereby a different but indistinguishable atom. They rush together violently, and a portion of the energy of their collision is given off in the form of light radiation. We conclude, therefore, from these laboratory determinations, based on many ingenious experiments, that the light along the path of an electric discharge is not owing immediately to the passage of the electricity, but to the quick automatic repair of the atoms it shattered along its course.

But whatever the process in detail of a lightning discharge within a cloud, or between a cloud and the earth, it is evident that only an exceptionally speedy device can adequately record it, for it is indeed "quick as lightning"—all over, a relativist might say, before it begins. Such a record is made photographically. The whole camera, for instance, is turned about a vertical axis at a known speed; or the light delivered to the camera lens by a rapidly rotating mirror; or two duplicate lenses equally distant from, and on opposite sides of, a common centre of rotation are rapidly spun about this centre in front of a stationary photographic plate; or the same, except that the lenses are stationary and the plate rotated; or by any equivalent device that will separate the images on the plate of closely sequent discharges, and show the time intervals between them when the rate of angular rotation of the moving element is given and focal lenght of the lens is known. It is fine sport, and several enthusiasts, playing the game most skillfully, have won at it a number of rich rewards.

In this way it has been learned that an electric discharge between a cloud and the earth commonly starts, like unto a small high speed projectile, from a negative lower portion of the cloud, over a devious course which it follows towards the earth at a uniform velocity around 125 miles per second, and along which so many wrecked atoms are left strewn as to render it electrically conducting. This lone pilot discharge is quickly followed and overtaken by rush after rush of electricity of the same sort traveling at the speed of roughly 12,000 miles per second, along the same path.

When the pilot is overtaken its speedy follower stops while it itself keeps on, but only to be overhauled again and again by a subsequent rush every time it has gotten ahead 33 feet, or so. At the same time fitful spurts of the opposite kind of electricity reach up from the earth until finally, that is, in the fiftieth of a second, perhaps, a conducting channel has been established from cloud to earth, whereupon there is a mighty rush of electricity up the prepared channel at the enormous velocity of roughly 25,000 miles per second (around the world in a second of time) and out any side branches that may have split off from it in the course of its development and extension from cloud to earth. This final, or return, rush affords much the brightest part of the entire lightning flash, and itself is brightest near the earth, and fades off with increase of distance along its course.

In many cases a lightning discharge is immediately followed by one or more others, as the flickering of its light indicates, generally at irregular intervals of a few hundredths of a second and along the same main course. These sequent discharges do not contain any of the intermediary pulsations, or gushes, only the initial pilot and the final return surge. Furthermore, only the first one or two of the sequent discharges, and commonly not even they, show any branches, and such as they do show are along the courses of the branches from the first discharge; and while the duration of any individual stroke is of the order of only one one-hundredth of a second, that of a sequence group may be anything up to an entire second, or even somewhat longer.

When the discharge is between cloud and cloud, or just from cloud to the space roundabout, it seems merely to diffuse along its course, or branched courses, into thin air until it, pilot and all, is too feeble to be perceptible. Neither, in this case, is there any return stroke, so far as we know—certainly nothing like the brilliant return sent up from the earth.

The lightning above described generally is called *streak lightning*, for the obvious reason that it occurs in streaks; not straight, nor zigzag with sharp angles and straight links, as artists used to represent it, but crooked with rounded or gradual turns. It also frequently is called *forked lightning*, in allusion, of course, to its appearance when divided into two or more branches. Another common name for it is *chain lightning*, having reference to its many apparent lines or segments. Sometimes a streak of lightning has seemed to lengthen at a perceptible speed instead of appearing to be all at once with no part first. Such real, or seeming, discharges have been called *rocket lightning*. It has not often been reported, and indeed may be merely an illusion.

On rather rare occasions a lightning streak persists for half a second, or longer like a suspended glowing wire, and then on fading away seems to divide into many bright and dark segments. The bright parts evidently are those viewed more or less end-on by the observer, since to him the light from the whole of each such portion appears to come from nearly the same spot, which spot therefore to him is relatively bright; while the dark (less bright) segments, on the other hand, are the portions of the streak which are rather across the observer's vision—segments whose light reaches the observer diffusely from a line and not concentratedly as from a point. Naturally, therefore, such streaks might be called, and they are called, *beaded lightning* and *pearl lightning*.

A far more frequent apparent form of lightning is the beautiful glow, momentary or briefly flickering, that often is seen of a summer's night in a distant cumulus cloud. This is not a special and distinct form of lightning as often has been assumed, and to which the names *sheet lightning* and *heat lightning* have been given, but just the diffusely reflected light from a streak invisible to the observer.

All the above forms of lightning, then, reduce to but one, streak lightning, and there remains just a single additional form to consider, namely, *ball liqhtning*—a humbug. Most cases of ball lightning are just after images, due to persistency of vision, from an instantaneous brilliant flash caught by the eye, similar to the after image one has of the sun following a momentary unprotected glance at it. One man reports find- ing this explanation for himself years after seeing ball lightning pass so close to him that he "felt the heat from it," but to which a cow he was milking at the time gave no attention. Of course not, as he himself now says, since the whole thing was in his own eye, and not an objective reality which the cow could experience. Another thing often called ball lightning is just a brush discharge, common on high peaks especially, during thunderstorm weather. Several other phenomena also have occasionally been reported as ball lightning, but not one of them has stood the test of careful examination.

All lightning then appears to be streak lightning, and every one wants to know how long the streak is, or can be. If we decide that brush discharges and tiny sparks, wherever they occur and however induced, are not lightning, or that even if they are they are not included under this query, but only streaks that may be seen at a considerable distance, then the answer is: Any length from perhaps 100 yards up to at least 12 miles, in the case of a meandering streak, stretched out to its full length.

It is natural, too, to want to know how hot the air is along the lightning's path, but this question cannot be answered in terms of definite figures on a thermometer scale. We can only say that some of the things that happen incident to the discharge, such as the explosion-like expansion that causes thunder, are just what one would expect from a suddenly produced high temperature. In short its effective temperature is very great.

Well, if the temperature is exceedingly high along a streak of lightning, doesn't the air there get scorched? Yes, in a sense, and many people say they can smell the scorched, or burnt, air and that always the odor is like that of burnt sulphur. The fact is that the lightning disrupts many of the oxygen, nitrogen and water vapour molecules along its course and that in their resultant combination with whatever is present several substances are formed that were not there before, especially combinations of nitrogen with hydrogen and with oxygen. In the presence of water vapour, always in the air, the first of these goes over into ammonia, and the second into nitrous and nitric acids.

All these things, ammonia, the oxides of nitrogen and the nitrogen acids, are pungent, and therefore are mistaken by most people who are not familiar with them for the highly pungent fumes of burning sulphur with which they, like the rest of us, are familiar.

It is interesting to note here that this ammonia promotes plant growth, and that both the nitrogen acids, on reaching the ground, produce compounds with substances in the soil which are excellent plant food, from which facts one correctly infers that thunderstorms above enrich the soil beneath. However, thunderstorm frequency and soil fertility do not in general run, even approximately, parallel to each other since there are many other factors that contribute to the final result. Lightning only makes the soil richer than it otherwise would be.

Lightning however, can do more things than furnish rich food for plants and make gloriously beautiful an otherwise dark and somber cloud. It can rend, crush, explode, fuse, and fire objects that it actually his-in the twinkling of an eye tear to slivers a giant tree that for centuries had but grown bigger and stronger with every stress and strain wind and rain and storm did bring. Naturally we wonder how such sudden and complete destruction is effected. The answer is: By an explosion.

When the current only follows the grain of the wood along a narrow strip just under the bark, the damage often is rather slight—just the bark over that strip ripped off and the solid wood beneath only slightly slivered—but when a heavy discharge follows the centre of the tree, as it sometimes does, the result may be the collapse of the entire trunk and the scattering of its parts hundreds of feet away in every direction. Here along the course of the discharge whatever poorly conducting substances are actually encountered are, in large measure, not only volatilized but also ionized. Hence the volume of these substances, sometimes along the very heart of a tree, is made many fold greater with all the suddenness and violence, and therefore, destructive effects, of exploding dynamite, that is, dynamite suddenly reverting to its constituents in the gaseous form, and requiring forthwith a greatly larger volume.

If the object encountered is a conductor, anyone of three things may happen, depending on its size and nature. A fine wire, for instance, is completely volatilized. It may even be blown as a copper gas, say, through an insulating cloth jacket without scorch or other injury, leaving an apparent insulated wire that is all hollow insulation and nothing more, the wire having "spurlos" (without trace) vanished. A wire of medium size, however, may just fuse, or only get quite hot, and, in either case, scorch its insulating jacket, if so covered. Finally, a metal rod is likely to be only warmed more or less, except at the points of entrance and exit where there is certain to be some fusion and volatilization.

On rare occasions lightning produces an effect just the reverse of an explosion—a collapse. This is conspicuous when a metal (copper, say) tube is struck with a current just sufficient to render its wall plastic, in which case the tube is crushed in as if by an explosion all around it. Evidently there is some reactional push in as a result of the sudden expansion of the abruptly heated surrounding air, but the air within the tube will, for the same reason, push out quite as hard, if not harder. Hence the collapse of the tube must be owing to some other cause, and that cause has been found to be a magnetic squeeze between all portions of the wall that are simultaneously carrying a current of the same sort in the same direction.

This conclusion follows from the fact that a wire carrying a current is, so long as the current lasts, surrounded by a magnetic field of such nature that two such wires parallel to each other draw as close together as possible, and that a bundle of such wires pull themselves into a compact strand. If now we regard the wall of our tube as consisting of a bundle of wires, all simultaneously carrying currents of the same sort in the same direction, it will be clear that there must be a radial compression on it tending to crush the wall into an axial rod, and that it will be so crushed as soon as rendered quite plastic, as sometimes happens incident to heating by the strong current, 20,000 amperes, or more of a lightning discharge.

Since lightning can smash a huge tree to bits with the violence and suddenness of a terrific explosion, and do many other things that require herculean force to perform, it is not surprising that many practical people have insisted that lightning no longer be permitted to just look pretty and make a noise, but, like Niagara, harnessed to some prosaic machine and made to do menial work. Obviously, if one should undertake to obtain financial backing for such an undertaking as this, he would do well to have some idea as to how much energy there is in the average lightning discharge.

From the strength of the current, say 20,000 amperes, the electric push, or voltage, between cloud and earth, around 25,000 volts per inch, and the total duration of the discharge, say one one-thousandth of a second, we find that if the streak is half a mile long, the energy used up is the equivalent of about 2,500 horsepower continuously for 24 hours, and as there are, over the earth as a whole, some 44,000 flashes of lightning per day, we may divide our result by 10, to include minor discharges and provide a large margin of safety, and still have for the world's average total lightning energy the equivalent of fully 1,100,000,000 horsepower, continuously, day and night, year in and year out.

At the low rate of one cent per kilowatt hour this would come, in round numbers, to $200,000,000 per day. Easy calculation, this, and tempting to the man of means; but the electrical engineer will find his part of the job something like building a bridge to the moon. Nevertheless, the facts in the case are of the kind to incite many a pleasing fancy, and to make scientifically plausible endless alleged happenings more amazing and spectacular than ever Aladdin could effect, rub his lamp and his ring as he might.

It was stated above that there really is only one sort of lightning, bright streak lightning, and yet we frequently are asked to explain dark, or black, lightning, and if we venture to insist that there isn't any such thing we are promptly shown a picture that is presumed to put us in the awkward dilemma of admitting either inexcusable ignorance or wilful prevarication.

We don't have to admit either alternative, however, for the dark streaks on the picture where one would expect white ones, are the result of a curious photographic reversal, as it is called. When an exposure to one lightning discharge is followed by diffuse light from others, such as normally would "fog" the plate, the fainter initial streaks are reversed on the negative and therefore print black, while the heavier portions of the main streak image are reversed only along their borders and therefore print as white lines with dark edges. No, lightning itself is not dark, but dazzlingly bright, even though circumstances at the time may cause its picture to show up dark.

The above various effects and phenomena of lightning might seem to pretty thoroughly cover the field, but it has also a noisy by-product, thunder, that likewise clamors for a hearing. As previously explained, myriads of molecules all along a lightning discharge are torn apart into their constituent atoms, and vast numbers of the atoms stripped each of one or more of its electrons, everyone of which parts, atom, stripped atom and electron requires as much space as would the original whole, or molecule, at the same temperature and pressure-and the effective temperature is now very high.

Hence from end to end of the lightning path there is an abrupt and violent explosion, so sudden that the surrounding air cannot just flow away gently, but only crowd and compress without quarter. Each crowded sheath around the path taken by the discharge, passes the push, with its consequent crowding and compression, on to the next, the next and the next until the volume affected becomes so large that the compression is negligible and the noise (thunder) inaudible. The distance at which this occurs, that is, the distance to which thunder can be heard, necessarily varies greatly with the course of the stroke, whether cloud to earth (loudest), cloud to cloud or open space; volume of the discharge; height of streak above sea level (density of the air in which it occurs); direction and strength of the local winds, and anything else that can affect the carry of sound. However, it seldom is heard more than ten to twelve miles from its source, which, as we know, is not a fourth the distance cannons often are heard.

This difference is owing mainly to the fact that all the cannon's noise comes from one place, the mouth of the cannon, while that of the lightning is from a crooked line a mile long or more, only the nearest portion of which gives the first, and, normally, loudest crash. That is, we have the benefit of the cannon's total effort, but of only a small fraction, at any one instant, of that of the lightning.

Another reason why a cannon often is heard surprisingly farther than thunder is the fact that the cannon may be fired when the air is still, and even in the early morning when the lower air is cool and therefore in its best condition for carrying sound, while thunder occurs only when the air is turbulent, a condition that causes it to scatter and diffuse sound much as crinkled glass wastes light.

The cannon, however, gives but a single boom, while thunder keeps on rumbling and roaring for many seconds, even nearly an entire minute in some cases, owing to the fact that the distance to the farthest portion of the lightning streak often is several miles (it takes sound five seconds to travel a mile) greater than that to its closest point. Clearly, too, when any appreciable portion of the streak is throughout at about the same distance from one he hears the noise from all of that portion simultaneously, and therefore as a loud boom. This is the main cause of the rumbling of thunder. In favorable localities there is more or less echoing of thunder by cliffs and other obstacles, but this effect commonly is of small importance. Especially is the echoing of thunder by clouds, so often alleged, practically negligible, as the mathematical analysis of the subject conclusively proves, and as anyone can easily verify by yelling at a dense fog bank and listening for the echo, an echo that never comes, unless from a house, cliff, or other solid object, in the fog.

It may, not seem reasonable, but it is so all the same: An occasional close-by stroke of lightning plays a tune for us, even if it does scare us, at least one note of what might be a tune, hence the name *musical thunder*. It comes by its music this way: When a number of sequent discharges, or strokes, follow each other, as they often do, at intervals of only two or three hundredths of a second, their separate booms or bangs blend into a musical note of a pitch determined by this frequency. At any considerable distance from the lightning the musical quality of the thunder is lost in the general rumble.

HAZARD AND PROTECTION

Although lightning is surpassingly beautiful, at a distance; its deep rumbly voice pleasant o hear, afar off; and even its music interesting, when it is over; nevertheless we must not grow contemptuous of it, nor play with it without adequate protection, for despite its allurements it really is not the mildest and safest thing in the world. True it is not one of our major hazards, but, for all that, hazard enough to command respect—400 lives and 1,000 injuries a year in the United States alone, and around $20,000,000 in property damage. However, it is quite practicable to insure fair to excellent protection from this danger. An allmetal house, for instance, would afford perfect immunity.

A metal-frame house also affords good protection, provided the lower portions of the metal parts are suitably "grounded," that is, connected by very heavy wires (copper is best) to water pipes, or to a good sized metal sheet (preferably copper) several feet down in the earth where the soil never gets very dry. A metal roof with every corner well grounded likewise affords good protection from lightning. Other houses, those that are neither metal, metal framed, nor metal roofed, can be rendered reasonably lightning proof only by means of a properly installed system of lightning rods. Just what constitutes such a system is a story too tedious to tell here. Anyone bent on putting up his own rods can get an excellent pamphlet on how to do it from either the National Bureau of Standards, Washington,3 D. C., or the U. S. Weather Bureau, Washington, D. C. However, in most cases it would be much better to do as the medical student, answering a quiz, said he would do for his patient, send for a doctor—turn the job over to a competent firm in that business.

No system of lightning protection prevents the occurrence of lightning, but it does do what is just as good, conveys the discharge between cloud and earth through itself and thereby away from everything that could be set on fire or otherwise injured. If one is caught in a thunderstorm then what? Come in out of the rain, if we can, or get in a canyon, or under a cliff, or in a cave (best of all); and in any case keep away from isolated trees in open spaces, avoid the crest of a hill, and never get closer than 100 feet, at least, to any wire fence, as we might be struck by lightning that hit the fence half a mile away.

Chapter 13

Air Pollution and Meteorology

Point Source Emissions

Point sources include industrial and nonindustrial stationary equipment or processes considered significant sources of air pollution emissions. A facility is considered to have significant emissions if it emits about one ton or more in a calendar year. Examples of point sources include industrial and commercial boilers, electric-utility boilers, turbine engines, wood and pulp processers, paper mills, industrial surface coating facilities, refinery and chemical processing operations, and petroleum storage tanks. Area sources that may fall under the "point source" definition are piping leaks, industrial wastewater treatment ponds, rock and quarry operations, and tank farms. Insignificant point sources are included by category in the area source inventory.

The most accurate method for determining the amount of emissions produced by a facility is with continuous emissions monitoring. However, this is expensive and not always practicable. Emissions are estimated using a variety of methods. Standard methods of estimating point source emissions are available from the U.S. Environmental Protection Agency in the *Compilation of Air Pollutant Emission Factors, Volume I*, also known as AP-42. Other methods include stack testing, equipment vendor test data, material balances such as for surface coating, or TCEQ-approved permit factors. If no other method is available, then the best engineering judgment must be used.

Facilities report point source emissions to the TCEQ for the calendar year and the data are stored in the Point Source Database. Data are available for use by TCEQ staff, the EPA, state and federal legislators, air pollution researchers, public interest groups and the general public.

EMISSIONS FROM NON-POINT SOURCES

On-Road Mobile Sources

Emissions from on-road mobile sources are estimated using a

sophisticated model called MOBILE, which was developed by the EPA. MOBILE calculates an emissions factor for mobile sources using a set of complex mathematical equations that require several user input values.

Vehicles are segregrated into eight vehicle classes, with MOBILE generating an emissions factor for each class and a composite emissions factor representing all classes.

On-Road Mobile Source Vehicle Classes

- Light-duty gasoline vehicles (LDGV)
- Light-duty gasoline trucks up to 6,000 pounds gross vehicle weight (LDGT1)
- Light-duty gasoline trucks from 6,001 to 8,500 pounds gross vehicle weight (LDGT2)
- Heavy-duty gasoline vehicles with more than 8,500 pounds gross vehicle weight (HDGV)
- Light-duty diesel vehicles (LDDV)
- Heavy-duty diesel vehicles with more than 8,500 pounds gross vehicle weight (HDDV)
- Motorcycles (MC)
- Light-duty diesel-powered trucks (LDDI)

After an emissions factor is generated for each vehicle class, the factor is then used in conjunction with the vehicle miles traveled (VMT) estimates, which were developed with the Texas Highway Performance Monitoring System (HPMS) data set for that selected area. This combination determines the contribution of emissions from mobile sources in a city, county, or state. VMT data is maintained by the Texas Department of Transportation.

The HPMS Programme produces VMT estimates for various roadway types categorized into two major population areas—urban and rural.

Roadway Types	
Urban	Rural
Interstate	Interstate
Other freeways and expressways	Other principal arterial
Other principal arterial	Minor arterial
Minor arterial	Major collector
Collector	Minor collector
Local	Local

In Texas, emissions from mobile sources are estimated on a county-wide basis. With a few exceptions nationwide, on-road mobile sources constitute the largest single source category of air pollution.

Non-Road Mobile Sources

Nonroad mobile sources include a wide variety of internal combustion engines not associated with highway vehicles. Emissions calculation methodology is as varied as the categories themselves. A federal computer model using engine types and landing-takeoff cycles is used to calculate most aircraft emissions. Actual fuel usage and track mileage are applied to determine locomotive emissions. Data on ship and barge traffic is used to calculate emissions from ocean vessels.

Nonroad Mobile Source Categories:

- *Aircraft:* Commercial Military
- *General:* LocomotivesVessels (ships and barges)
- *Small Engines:* Lawn and garden Airport support vehicles Recreational marineLight commercialIndustrial construction Agricultural logging.

The EPA developed emissions factors for a lengthy list of small engine categories, which included research on individual engines and fuel types. Included in this category are two major emissions sources—lawn mowers and recreational boat engines.

Area Sources

Area source inventories generally report emissions by categories rather than by individual source; a common method in reporting point source emissions. Area source emissions are calculated by various methods and depend on the type of data available for each category. For example, whenever fuel use and materials data are not available, employee and county population numbers are used with established emission factors to calculate emissions. Emissions are calculated and reported on a county-wide basis.

Major categories of area sources are:

- Stationary source fuel combustion such as residential fuel combustion
- Solvent use (e.g., small surface coating operations)
- Product storage and transport distribution (e.g., gasoline)
- Light industrial/commercial sources
- Agriculture (e.g., feedlots, crop burning)
- Waste management (e.g., landfills)
- Miscellaneous area sources (e.g., forest fires, wind erosion, unpaved roads)

Biogenic Sources

Biogenic emissions account for 30 per cent of all the volatile organic compounds (VOCs) emitted in urban areas in the eastern half of Texas. For the purposes of photochemical modeling, biogenic VOC emissions are estimated using a computer model that takes into account the species of trees present, the density of their foliage, the temperature and solar radiation on the day in question, and the distribution of vegetation throughout the modeling domain. It is important to measure these parameters accurately if the biogenics inventory is to be correct. The TCEQ has hired specialists to measure some of these variables in north-central and southeastern Texas.

Most plants emit some VOCs, but the largest emitters are oaks, pines, sweet gums, eucalypti, and poplars. Some VOCs are easily detected by their aroma. Pines, sycamores, and eucalypti emit fragrant monoterpenes, while other VOCs—such as isoprene—are not as aromatic.

Isoprene is a byproduct of photosynthesis. Scientists still debate the purpose of its emission but some evidence suggests that plants can cope better with heat if isoprene is present. Because it is generated by photosynthesis, isoprene emissions are not generated at night.

Monoterpenes are known as "essential oils." There is solid evidence that plants make monoterpenes, which are found in small reservoirs in the leaves or needles of plants, to ward off herbivores. When an insect feeds on the leaf, the monoterpenes are released and can adversely affect the insect's health. Because the monoterpenes are always present in the leaves, their emission rate depends mostly on the temperature. Higher temperatures will evaporate larger amounts into the atmosphere.

There are a few other important organic compounds emitted by plants. Alcohols are often emitted by damaged vegetation; there is some evidence that these alcohols act as an antiseptic. A few recent studies suggest that alkenes are also emitted by some plants.

COMBUSTION

Combustion is the act or process of burning. For combustion to occur, fuel, oxygen (air), and heat must be present together.

Per definition combustion is the chemical reaction of a particular substance with an oxidant. Generally this will mean atmospheric oxygen and will be treated as such in the rest of this site.

The combustion process is started by heating the fuel above its ignition temperature in the presence of oxygen. Under the influence of heat, the chemical bonds of the fuel are split. If complete combustion takes place, the elements carbon (C), hydrogen (H) and sulphur (S) react with the oxygen content of the air to form carbon dioxide CO_2, water vapour H_2O and sulphur dioxide SO_2 and, to a lesser degree, sulphur trioxide SO_3.

If not enough oxygen is present or the fuel / air mixture is insufficient then the burning gases are partially cooled below the ignition temperature (too much air or cold burner walls), and the combustion process stays incomplete. The flue gases then still contain burnable components, mainly carbon monoxide CO, carbon C (soot) and various hydrocarbons C_xH_y. Since these components are, along with NO_x, pollutants which harm our environment, measures have to be taken to prevent the formation of them.

To ensure complete combustion, it is essential to provide a certain amount of excess air. Combustion optimisation saves money!

The quality of a combustion system is determined by a maximum percentage of complete combustion, along with a minimum of excess air (commonly 5 to 20% above the necessary level for ideal combustion)

AIR	+	FUEL	>>>	FLUE GAS
Oxygen (O_2)		Carbon (C)		Carbon dioxide (CO_2)
		Hydrogen (H_2)		Carbon monoxide (CO)
Nitrogen (N_2)		Sulphur (S)		Sulphur dioxide (SO_2)
		Oxygen (O_2)		excess O_2 (NO_2)
Water vapour		Nitrogen (N_2)		Nitrogen oxides
		Water (H_2O)		Nitrogen (N_2)
				Water vapour
				Soot

Flue gas will generally contain a certain amount of CH_4 (methane) if the combustion was not complete. Other hydrocarbons will not occur under normal conditions. higher hydrocarbons are only produced under conditions of high pressure and high temperature such as occur in an internal combustion engine.

Combustion Efficiency

Combustion efficiency is a calculation of how well our equipment is burning a specific fuel, shown in per cent. Complete combustion efficiency would extract all the energy available in the fuel. However 100% combustion efficiency is not realistically achievable. Common combustion processes produce efficiencies from 10% to 95%.

Combustion efficiency calculations assume complete fuel combustion and are based on three factors:

- The chemistry of the fuel.
- The net temperature of the stack gases.
- The percentage of oxygen or CO_2 by volume after combustion.

Combustion processes and their combustion efficiency ranges

Process	Typical combustion efficiency range
Home fireplace	10-30 %
Space heater	50-80 %
Commercial gas boiler	70-82 %
Residential gas boiler with atmospheric burner "Low efficiency"	70-82 %
Oil burner heating system	73-85 %
induced draft furnace "medium efficiency"	74-80 %
boiler with gas burner	75-85 %
Condensing furnace (Gas & Oil) "High efficiency"	85-93 %

If our calculation shows that our equipment is losing 20 % of the heating energy of the fuel through stack losses, our equipment is running at 80 % efficiency.

Combustion efficiency relates to the part of the reactants that combine chemically. Combustion efficiency increases with increasing temperature of the reactants, increasing time that the reactants are in contact, increasing vapour pressures, increasing surface areas and increasing stored chemical energy. One way of increasing the temperature of the reactants and their vapour pressures is to preheat them by circulating them around the combustion chamber and throat before being injected into the combustion chamber. The specific heat of combustion is a chemical property that refers to the amount of energy that can theoretically be extracted from a fuel at 100 % combustion efficiency. The heating value is a more realistic term and does not include the condensation of the water vapour produced. It is thus more easily applied to real combustion processes.

Air preheating is one method used in steel works, for instance, to increase combustion efficiency. This uses the heat in the flue gases to heat one of a pair of chambers and the inlet air passes through the other one. The use of the chambers is switched as soon as the one chamber has reached temperature, so the air passes through the heated chamber. This is one of the simplest and best methods of increasing combustion efficiency in this kind of process. Such preheaters are standard euipment these days for larger systems.

The introduction of the condensing burner has led to the strange situation where combustion efficiencies in excess of 100 % are reported. These devices also use the specific heat of vaporisation as a surce of energy and therefore have an increased yield, provided they are operated in the appropriate temperature range.

Measurement of Combustion Efficiency

Assuming that the fuel parameters are known, only the oxygen or carbon dioxide concentration, ambient temperature and flue gas temperature have to be determined. Since that does not give a 100% certain reading, it is really necessary to measure CO as well.

Combustion efficiency is a simple (and occasionally misleading) measure of the heating efficiency of a boiler. It is equal to 100 per cent minus the percentage of heat lost up the vent (called "flue loss" or "stack loss").

The formal methods for measuring flue loss vary by furnace or boiler type, but, in essence, the combustion efficiency measures the total heat energy that is spread through the heating system or escapes from the boiler jacket itself (jacket loss) instead of going up the stack or flue to warm up the environment. Certain factors may place a limit on the combustion efficiency that can be reached. In particular, the need to keep the gases above the dewpoint for sulfur dioxide may be a major factor. Such considerations must be borne in mind at all times when adjusting a burner system.

Combustion efficiency is based on the flue gas temperature and inlet air temperature. These must be measured in an appropriate fashion. The inlet air at the air inlet, avoiding any wind chill factor that the air movement might produce. This is a relatively simple, but important factor. The flue gas temperature must be measured in the hottest part of the gas stream. Generally this will be in the centre of the flue, but eddy effects may cause it to be slightly displaced. It is vital to have the correct fuel programmed into the flue gas analyser before measuring combustion efficiency. The fuel parameters have a noticeable effect on the value of combustion efficiency, and failure to follow this step will result in erroneous data.

Combustion efficiency measures the so-called "steady-state" efficiency of a boiler, which may or may not be appropriate in specific cases, especially where a burner is used as back-up and only ever operates for short periods of time to cover peak usage. many burners operate on a short cycle which is intrinsically less efficient than continuous use at full power.

As stated above, combustion efficiency may be limited in certain cases and with the condensing boilers it is a definite problem. Here the quantity of condensate produced in a set time must be measured for any calculation of efficiency. The high specific heat of evaporation for water makes this a major factor in any true calculation of efficiency. Nevertheless, it is a useful factor for seeing changes made by adjustments. Generally this data is also required when annual checks are made on equipment.

INCINERATION

Incineration is a waste treatment technology that involves the combustion of organic materials and/or substances. Incineration and other high temperature waste treatment systems are described as "thermal treatment". Incineration of waste materials converts the waste into incinerator bottom ash, flue gases, particulates, and heat, which can in turn be used to generate electric power. The flue gases are cleaned of pollutants before they are dispersed in the atmosphere.

Incineration with energy recovery is one of several waste-to-energy (WtE) technologies such as gasification, Plasma arc gasification, pyrolysis and anaerobic digestion. Incineration may also be implemented without energy and materials recovery.

In several countries there are still expert and local community concerns about the environmental impact of incinerators.

In some countries, incinerators built just a few decades ago often did not include a materials separation to remove hazardous, bulky or recyclable materials before combustion. These facilities tended to risk the health of the plant workers and the local environment due to inadequate levels of gas cleaning and combustion process control. Most of these facilities did not generate electricity.

Incinerators reduce the volume of the original waste by 95-96 %, depending upon composition and degree of recovery of materials such as metals from the ash for recycling. This means that while incineration does not completely replace landfilling, it reduces the necessary volume for disposal significantly.

Incineration has particularly strong benefits for the treatment of certain waste types in niche areas such as clinical wastes and certain hazardous wastes where pathogens and toxins can be destroyed by high temperatures. Examples include chemical multi-product plants with diverse toxic or very toxic wastewater streams, which cannot be routed to a conventional wastewater treatment plant.

Waste combustion is particularly popular in countries such as Japan where land is a scarce resource. Denmark and Sweden have been leaders in using the energy generated from incineration for more than a century, in localised combined heat and power facilities supporting district heating schemes.

Pollution

Incineration has a number of outputs such as the ash and the emission to the atmosphere of flue gas. Before the flue gas cleaning system, the flue gases may contain significant amounts of particulate matter, heavy metals, dioxins, furans, sulfur dioxide, and hydrochloric acid.

In a study from 1994, Delaware Solid Waste Authority found that, for same amount of produced energy, incineration plants emitted fewer particles, hydrocarbons and less SO_2, HCl, CO and NO_x than coal-fired power plants, but more than natural gas fired power plants. According to Germany's Ministry of the Environment, waste incinerators reduce the amount of some atmospheric pollutants by substituting power produced by coal-fired plants with power from waste-fired plants

Gaseous Emissions

Dioxin and Furans

The most publicized concerns from environmentalists about the incineration of municipal solid wastes (MSW) involve the fear that it produces significant amounts of dioxin and furan emissions. Dioxins and furans are considered by many to be serious health hazards. Older generation incinerators that were not equipped with adequate gas cleaning technologies were indeed significant sources of dioxin emissions. Today, however, due to advances in emission control designs and stringent new governmental regulations, incinerators emit virtually no dioxins. In 2005, The Ministry of the Environment of Germany, where there were 66 incinerators at that time, estimated that "...whereas in 1990 one third of all dioxin emissions in Germany came from incineration plants, for the year 2000 the figure was less than 1 %. Chimneys and tiled stoves in private households alone discharge approximately twenty times more dioxin into the environment than incineration plants.". According to the United States Environmental Protection Agency, incineration plants are no longer significant sources of dioxins and furans. In 1987, before the governmental regulations required the use of emission controls, there was a total of 10,000 grams (350 oz) of dioxin emissions from U.S. incinerators. Today, the total emissions from the 87 plants are only 10 grams (0.35 oz) yearly, a reduction of 99.9 %. Backyard barrel burning of household and garden wastes, still allowed in some rural areas, generates 580 grams (20 oz) of dioxins yearly. Studies conducted by EPA demonstrate that the emissions from just one family using a burn barrel produces more emissions than an incineration plant disposing of 200 metric tons (220 short tons) of waste per day.

Generally the breakdown of dioxin requires exposure of the molecule to a sufficiently high temperature so as to trigger thermal breakdown of the molecular bonds holding it together. When burning of plastics outdoors in a burn barrel or garbage pit such temperatures are not reached, causing high dioxin emissions as mentioned above. While the plastic does burn in an open-air fire, the dioxins remain after combustion and float off into the atmosphere.

Modern municipal incinerator designs include a high temperature zone, where the flue gas is ensured to sustain a temperature above 850 °C for at least 2 seconds befores it is cooled down. They are equipped with auxiliary heaters to ensure this at all times. These are often fueled by oil, and normally only active for a very small fraction of the time. A side effect controlling dioxin is the potential for generation of reactive oxides (NO_x) in the flue gas, which must be removed with SCR or SNCR.

For very small municipal incinerators, the required temperature for thermal breakdown of dioxin may be reached using a high-temperature electrical heating element, plus an SCR stage.

CO_2

As for other complete combustion processes, nearly all of the carbon content in the waste is emitted as CO_2 to the atmosphere. MSW contain approximately the same mass fraction of carbon as CO_2 itself (27%), so incineration of 1 metric ton (1.1 short tons) of MSW produce approximately 1 metric ton (1.1 short tons) of CO_2.

In the event that the waste was landfilled, 1 metric ton (1.1 short tons) of MSW would produce approximately 62 cubic metres (2,200 cu ft) methane via the anaerobic decomposition of the biodegradable part of the waste. This amount of methane has more than twice the global warming potential than the 1 metric ton (1.1 short tons) of CO_2, which would have been produced by incineration. In some countries, large amounts of landfill gas are collected, but still the global warming potential of the landfill gas emitted to atmosphere in the US in 1999 was approximately 32 % higher than the amount of CO_2 that would have been emitted by incineration.

In addition, nearly all biodegradable waste has biological origin. This material has been formed by plants using atmospheric CO_2 typically within the last growing season. If these plants are regrown the CO_2 emitted from their combustion will be taken out from the atmosphere once more.

Such considerations are the main reason why several countries administrate incineration of the biodegradable part of waste as renewable energy. The rest - mainly plastics and other oil and gas derived products - is generally treated as non-renewables.

Different results for the CO_2 footprint of incineration can be reached with different assumptions. Local conditions (such as limited local district heating demand, no fossil fuel generated electricity to replace or high levels of aluminum in the waste stream) can decrease the CO_2 benefits of incineration. The methology and other assumptions may also influence the results significantly. For example the methane emissions from landfills occurring at a later date may be neglected or given less weight, or biodegradable waste may not be considered CO_2 neutral. A recent study by Eunomia Research and Consulting on potential waste treatment technologies in London demonstrated that by applying several of these unusual assumptions the average existing incineration plants performed poorly for CO_2 balance compared to the theoretical potential of other emerging waste treatment technologies.[

Other Emissions

Other gaseous toxins in the flue gas from incinerator furnaces include sulfur dioxide, hydrochloric acid, heavy metals and fine particles.

The steam content in the flue may produce visible fume from the stack, which can be perceived as a visual pollution. It may be avoided by decreasing the steam content by flue gas condensation and reheating, or by increasing the flue gas exit temperature well above its dew point. Flue gas condensation allows the latent heat of vaporization of the water to be recovered, subsequently increasing the thermal efficiency of the plant.

Flue Gas Cleaning

The quantity of pollutants in the flue gas from incineration plants is reduced by several processes.

Particulate is collected by particle filtration, most often electrostatic precipitators (ESP) and/or baghouse filters. The latter are generally very efficient for collecting fine particles. In an investigation by the Ministry of the Environment of Denmark in 2006, the average particulate emissions per energy content of incinerated waste from 16 Danish incinerators were below 2.02 g/GJ (grams per energy content of the incinerated waste).

Detailed measurements of fine particles with sizes below 2.5 micrometres ($PM_{2.5}$) were performed on three of the incinerators: One incinerator equipped with an ESP for particle filtration emitted 5.3 g/GJ fine particles, while two incinerators equipped with baghouse filters emitted 0.002 and 0.013 g/GJ $PM_{2.5}$. For ultra fine particles ($PM_{1.0}$), the numbers were 4.889 g/GJ $PM_{1.0}$ from the ESP plant, while emissions of 0.000 and 0.008 g/GJ $PM_{1.0}$ were measured from the plants equipped with baghouse filters.

Acid gas scrubbers are used to remove hydrochloric acid, nitric acid, hydrofluoric acid, mercury, lead and other heavy metals. Basic scrubbers remove sulfur dioxide, forming gypsum by reaction with lime.

Waste water from scrubbers must subsequently pass through a waste water treatment plant.

Sulfur dioxide may also be removed by dry desulfurisation by injection limestone slurry into the flue gas before the particle filtration.

NO_x is either reduced by catalytic reduction with ammonia in a catalytic converter (selective catalytic reduction, SCR) or by a high temperature reaction with ammonia in the furnace (selective non-catalytic reduction, SNCR).

Heavy metals are often adsorbed on injected active carbon powder, which is collected by the particle filtration.

Other Pollution Issues

Odor pollution can be a problem with old-style incinerators, but odors and dust are extremely well controlled in newer incineration plants. They receive and store the waste in an enclosed area with a negative pressure with the airflow being routed through the boiler which prevents unpleasant odors from escaping into the atmosphere. However, not all plants are implemented this way, resulting in inconveniences in the locality.

An issue that affects community relationships is the increased road traffic of waste collection vehicles to transport municipal waste to the incinerator. Due to this reason, most incinerators are located in industrial areas.

PHOTOCHEMICAL POLLUTION

Photochemical pollution (or photo-oxidising pollution) is a series of complex phenomena leading to the formation of ozone and other oxidising compounds (hydrogen peroxide, aldehydes, peroxy acetyl nitrate or PAN) from primary pollutants (known as precursors - oxides of nitrogen and volatile organic compounds or VOCs) and energy provided by ultra-violet (UV) radiation from the sun. These phenomena occur in the layers of air close to the ground, and in the free troposphere. The ozone formed at this level is called ground-level or "bad ozone" because of its damaging effect on human health and vegetation. Conversely, the ozone in the stratosphere (at an altitude of 19-30 km), is called "good ozone" because it protects us from solar UV radiation.

Surprisingly, the concentrations of ozone measured far from sources of precursors (built-up areas, for example) are higher than those measured near the sources. This is because, in a city for example, the emissions of NO (mainly from traffic) are high. Ozone is destroyed by NO. The NO acts as a sink of ozone because it uses it up. If the cloud of pollutants formed in the city moves to the countryside, or if emissions of NO are reduced, the concentration of ozone increases because the ozone is no longer consumed.

Photochemical pollution is a characteristic phenomenon of summer anticyclonic situations.

Ozone has damaging effects on human health, forest and agricultural ecosystems. In addition, this phenomenon of photo-oxidising pollution is closely linked to acid rain.

THE GREENHOUSE EFFECT

The greenhouse effect is a natural phenomenon involving the absorption of very long wave infra-red (IR) radiation, reflected from the surface of the earth, by compounds present in the atmosphere: CO_2, CH_4, H_2O, O_3, N_2O, CFCs. Part of the IR radiation is not reflected back into space. The result is energy absorption. This energy is transformed into heat. Most of these compounds are present in natural form, which has enabled the development and survival of life on Earth. The average temperature on earth is 15°C; if the natural greenhouse effect did not exist, the average temperature would be –18°C.

Since the industrial revolution, there has been an increase in the concentration of greenhouse gases:

- CO_2 mainly from industrial, domestic and transport combustion.
- CH_4 mainly from farming practices: e.g. rice growing, livestock.
- N_2O mainly from farming practices.
- CFCs (now banned), HFCs, PFCs, SF_6.

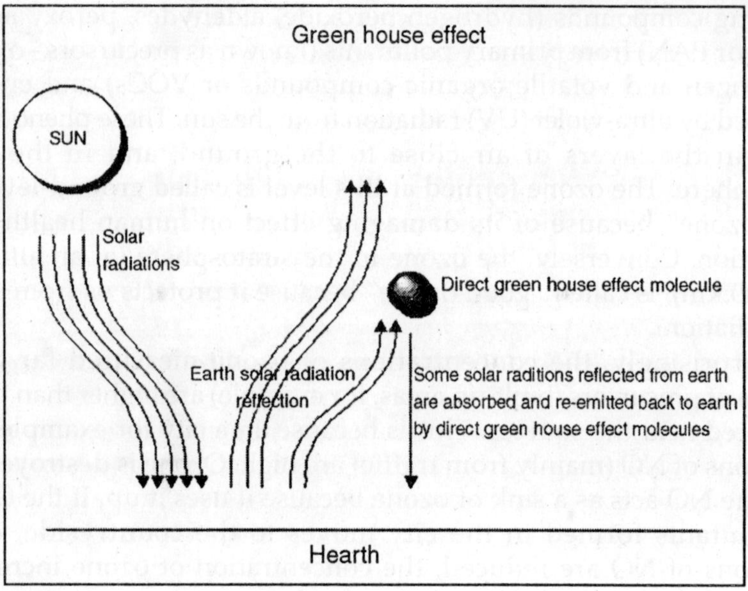

DEPLETION OF THE OZONE LAYER

Ozone is the main component in the upper atmosphere at an altitude of 25 km. The ozone layer is said to be "good ozone" because it absorbs UV radiation from the sun and so protects us from the risks of skin cancer and other genetic mutations. It also protects the photosynthesising activity of plants.

An abnormal lowering in concentrations of ozone at the South pole at the end of the polar winter, when the sun returns, was discovered in 1980. At the end of the southern winter, when the sun returns, the ozone content diminishes by 40 to 60%. The maximum deficiency is about 20 - 25 km above ground level.

Many compounds can destroy ozone (OH, H, NO, Cl, Br, HO_2). A high correlation was found between ozone deficiency and concentrations of ClO. The presence of the radicals Cl and ClO in the stratosphere is due to the natural emission of methylene chloride by oceans and to chlorofluorocarbons (CFCs) emitted by human activities. CFCs are very stable molecules. They are transported into the stratosphere where they release chlorine, thus disturbing the natural balance which governs the presence of ozone at this altitude.

The annual phenomenon of depleted concentration of ozone is more marked at the South pole than at the North pole because of the different conditions prevailing there. At the South pole, a vortex appears during the winter. The temperatures are around –80° to –100°C. Clouds then contain fine crystals of ice which fix the chlorine in the form of HCl and NO_2ClO. As soon as the sun returns, its UV radiation releases the radicals Cl and ClO which quickly react with the ozone. At the North pole no such vortex is formed. Instead, there are a multitude of small holes.

AIR POLLUTION AND METEOROLOGY

The science of meteorology has great bearing on air pollution. An air pollution problem involves three parts: the source, the movement of the pollutant and the recipient. All meteorological phenomena are a result of interaction of the elemental properties of the atmosphere, heat, pressure, wind and moisture. In this lecture we will discuss the meteorological conditions, which directly influence the transport and dispersion of pollutants.

Wind

Wind is simply air in motion. On global or macroscale wind patterns are set up due to unequal heating of earth surface by solar radiation at the equator and the polar regions, rotation of the earth and the difference between conductive capacities of land and ocean masses. Secondary or mesoscale circulation patterns develop because of the regional or local topography. Mountain ranges, cloud cover, waterbodies, deserts, forestation, etc., influence wind patterns on scales of a few hundred kilometers.

Accordingly a pattern of wind is setup, some seasonal and some permanent. Microscale phenomenon occurs over areas of less than 10 km extent. Standard wind patterns may deviate markedly due to varying frictional effects of the earth surface, such as, rural open land, irregular topography and urban development, effect of radiant heat from deserts and cities, effect of lakes, etc.

The movement of air at the mesoscale and microscale levels is of concern in control of air pollution. A study of air movement over relatively small geographical regions can help in understanding the movement of pollutants.

The variation of the horizontal wind speed with height is important in evaluating diffusion from stacks. Close to the ground the effects of friction retard the wind flow and cause it to change the direction as well. In the upper layers, 200 m to 500 m above ground, the wind speed reaches the maximum value.

It is obviously important in predicting pollutant dispersion to know the direction of wind.

The wind direction and speed data may be collected every hour in a month and classified according to speed and direction. It is then summarized in the form of a polar diagram called *wind rose*. The position of the spokes show the direction from which the wind was blowing, the length of various segments of the spokes show the per cent of time the wind was of the designated speed. Thus from the diagram, most often (12% of time) the wind was from SE; the strongest wind (9-11 m/s) was from NW and NNW.

Lapse Rate

As a parcel of air rises in the earth's atmosphere it experiences lower and lower pressure from the surrounding air molecules, and thus it expands. This expansion lowers its temperature. Ideally, if it does not absorb heat from its surroundings and it does not contain any moisture, it cools at a rate of 1ºC/100 m rise. This is known as *dry adiabatic lapse rate.* If the parcel moves down it warms up at the same rate.

For a particular place at a particular time, the existing temperature can be determined by sending up a balloon equipped with a thermometer. The balloon moves through the air, and not with it. The temperature profile of the air, which the balloon measures, is called the *ambient lapse rate, environmental lapse rate,* or the *prevailing lapse rate.*

A super-adiabatic lapse rate also called a strong lapse rate occurs when the atmosphere temperature drops more than 1ºC/100m. A sub-adiabatic rate also called weak lapse rate, is characterized by drop of less than 1ºC/100 m.

A special case of weak lapse rate is the inversion, a condition which has warmer layer above colder air.

During super-adiabatic lapse rate the atmospheric conditions are unstable. If a parcel of air at 500m elevation, at 20ºC is pushed upward to 1000m, its temperature will come down to 15ºC (according to adiabatic lapse rate).

The prevailing temparature is however 10ºC at 1000m. The parcel of air will be surrounded by colder air and therefore will keep moving up. Similarly if the parcel is displaced downwords, it will become colder than its surroundings and therefore will move down. Superadiabatic conditions are thus unstable, characterized by a great deal of vertical air movement and turbulence.

The sub-adiabatic condition is by contrast a very stable system. Consider again a parcel of air at 500 m elevation at 20ºC. If the parcel is displaced to 1000 m it will cool by 5ºC to 15ºC. But the surrounding air would be warmer. It will therefore fall back to its point of origin. Similarly if a parcel of air at 500 m is pushed down, it will become warmer than its surrounding and therefore will rise back to its original position. Thus such systems are characterized by very limited vertical mixing.

Inversion

An inversion is an extreme sub-adiabatic condition, and thus the vertical air movement within the inversion is almost nill. The two most common kind of inversion are *subsidence inversion* and *radiation inversion.*

The base of the subsidence inversion lies some distance above earth's surface. This type of inversion is formed due to adiabatic compression and warming of sinking air mass to a lower altitude in the region of a high pressure centre.

In the case of radiation inversion, the surface layers of the atmosphere during the day receive heat by conduction, convection and radiation from the earth's surface and are warmed. This results in a temperature profile in the lower atmosphere, which is represented by a negative temperature gradient. On a clear night, the ground surface radiates heat and quickly cools. The air layer adjacent to the earth surface are cooled to a temperature below that of the layers of air at higher elevations. This type of the inversion is strongest just before daylight when it may extend to 500 m. It breaks up as the morning sun heats the ground.

EFFECTS OF AIR POLLUTION

The air is shared among all living things. When it is polluted by a factory in Asia, a fire in Australia, a dust storm in Africa, or car emissions in North America, the sharing continues despite the fact that these chemicals and particles have detrimental effects.

Scientists have determined many of the harmful local effects of air pollution. We know, for instance, that air pollution can negatively impact human health and cause coughs, burning eyes, breathing problems, and even death. We know that atmospheric haze or smog reduces visibility and that acid rain from chemical emissions damages property, pollutes water resources, and can harm forests, wildlife, and agriculture.

But what are the regional and global impacts of air pollution? Through large scientific field campaigns such as MILAGRO, scientists are beginning to track its movement from cities into regional and global environments. Their goal is to determine air pollution's movement and impact on climate and atmospheric composition locally, regionally, and globally.

Is human-produced air pollution and its effects an example of the

"Tragedy of the Commons" — a concept that states that any resource open to everyone will eventually be destroyed? Despite the fact that people are creating much of today's air pollution, the answer will ultimately depend on how humankind responds to the problem. A lot has been done to improve air quality in recent decades, but we still have a long way to go.

AIR POLLUTION AND HUMAN HEALTH

People have no choice but to breathe the air around them. When it is polluted, they breathe in ozone, particles and harmful gases that can hurt their lungs, heart, and overall health. Air pollution can cause coughing, burning eyes, and breathing problems. Fortunately, people usually start to feel better as soon as the air quality improves, but not always.

63 people died in Belgium in 1930, 20 people died in Pennsylvania in 1948, and more than 4,000 died in London in 1952 as a result of severe air pollution. Breathing small amounts of air pollution over many years is also considered dangerous. It may even contribute to life-threatening diseases such as cancer.

The elderly, the young, and those with cardiopulmonary disease, such as asthma or severe bronchitis, are the most vulnerable to air pollution exposure. Children are at greater risk because their lungs are still growing. Also, they play outside and are active. As a result, pound for pound they breathe more outdoor air pollution than most adults.

Although people have no choice but to breathe the air around them, they do have choices that can help them stay healthy. They can choose to stay indoors or be less active on poor air quality days. They can avoid high-traffic and highly industrialized areas whenever possible. They can also choose to support collective efforts and take individual steps that reduce air pollution. Such actions are a positive response to a problem that can literally steal one's breath away.

AIR POLLUTION AND ATMOSPHERIC VISIBILITY

Have we ever spent time in a large city? If so, the odds are we've seen the sky engulfed in a brownish-yellow or grayish-white haze due to air pollution. Such haze can reduce visibility from miles (kilometers) to yards (meters). Mountains or buildings once in plain sight can suddenly be blocked from view.

Air pollution that reduces visibility is often called haze or smog. The term smog originally meant a mixture of smoke and fog in the air, but today it refers to any mixture of air pollutants that can be seen.

Smog typically starts in cities or areas with many people, but because it travels with the wind, it can appear in rural areas as well.

One consequence of smog over any given area is that it can change the area's climate. Certain dark particles, such as carbon, absorb solar radiation and scatter sunlight, helping produce the characteristic haze that fills the skies over the world's megacities. This haze reduces the amount of the Sun's energy reaching the Earth's surface, sometimes by as much as 35 per cent.

A reduction in sunlight may not be the only thing air pollution inhibits. Some research has supported the idea that certain air particles are altering rainfall patterns as well. Although particles in the air form the nucleus that attracts cloud moisture into water droplets, specks of soot or black carbon may be too small to produce raindrops big enough to hit the ground. Since rain flushes pollutants from the atmosphere, visibility could be negatively impacted as a consequence.

Scientific field campaigns such as MILAGRO are one way scientists can research atmospheric processes to prove or disprove such ideas. For now, if rain isn't in the forecast when atmospheric visibility is low and smog is high, than wind is likel our best hope for a return to clear skies.

AIR POLLUTION AND CLIMATE CHANGE

Factories Like This are a Major Source of Air Pollution

Air pollution changes our planet's climate, but not all types of air pollution have the same effect. There are many different types of air pollution. Some types cause global warming to speed up. Others cause global warming to slow down by creating a temporary cooling effect for a few days or weeks. Read on the learn more about the pollution that causes Earth to warm and the pollution that causes Earth to cool.

Some Air Pollutants cause more Global Warming

Air pollution includes greenhouses gases. One of these is carbon dioxide, a common part of the exhaust from cars and trucks. Greenhouse gases cause global warming by trapping heat from the Sun in the Earth's atmosphere. Greenhouse gases are a natural part of Earth's atmosphere, but in the last 150 years or so, the amount in our atmosphere has increased. The increase comes from car exhaust and pollutants released from smokestacks at factories and power plants. The increase in greenhouses gases is the cause of most of the global warming that happened over the past century. Scientists predict that much more warming will likely happen during the next century.

Some Air Pollutants cause Temporary Global Cooling

Cars, trucks, and smokestacks also release tiny particles into the atmosphere. These tiny particles are called aerosols. They can be made of different things such as mineral dust, sulfates, sea salt, or carbon. Some of these tiny particles block a little bit of the Sun's energy from getting to Earth. Some of these particles get into the atmosphere naturally. They are dust lifted into the atmosphere from deserts, from evaporating droplets from the ocean, released by the smoke from wildfires, and erupting volcanoes. But air pollution released by humans by burning of fossil fuels also adds them to the atmosphere.

Greenhouse gases stay in the atmosphere for years and cause warming around the world. Computer models indicate that, worldwide, the tiny aerosols cause about half as much cooling as greenhouse gases cause warming.

AIR POLLUTION TO PROPERTY

This building in Copola, Mexico, has been damaged by acid rain.

In addition to damaging the environment and human health, air pollution can harm buildings, monuments, outdoor statues, and other such structures. The chemicals in air pollution eat away at materials such as sandstone, limestone, mortar, and different metals. Acid rain dissolves stone and can create cracks in buildings.

Repairing this damage, particularly to historic structures, can be very expensive. The National Centre for Preservation Technology & Training studies the environmental effects of pollution on property and works to restore and protect historic structures and monuments.

Homeowners also pay a price for air pollution. Acid rain can dissolve paint and eat away at aluminum siding, while dirt particles in the air stick to a house and ruin its appearance.

AIR POLLUTION AFFECTS PLANTS, ANIMALS, AND ENVIRONMENTS

Both of these pots of clover plants have been growing for 30 days, but one looks healthier than the other! The clover plants in the top picture (A) were given normal water. The clover plants in the lower picture (B) were given acidic water (pH=2.0) for the latter 20 days. When air pollution causes acid rain, plants that rely on rain water to live and grow are endangered.

Some air pollutants harm plants and animals directly. Other pollutants harm the habitat, food or water that plants and animals need to survive. Read on to learn more about how air pollutants harm plants and animals.

Acid Rain Harms Living Things

When acidic air pollutants combine with water droplets in clouds, the water becomes acidic. When those droplets fall to the ground, the acid rain can damage the environment. Damage due to acid rain kills trees and harms animals, fish, and other wildlife. Acid rain can destroy the leaves of plants like in the picture at the left. When acid rain soaks into the ground, it can make the soil an unfit habitat for many living things. Acid rain also changes the chemistry of the water in lakes and streams, harming fish and other aquatic life.

The Thinning Ozone Layer Harms Living Things

Air pollutants called *chlorofluorocarbons*(or CFCs) have destroyed parts of the ozone layer.The ozone layer, located in the stratosphere layer of Earth's atmosphere, shields our planet from the Sun's ultraviolet radiation. The areas of thin ozone are called *ozone holes*. Ultraviolet radiation causes skin cancer and damages plants and wildlife.

Tropospheric Ozone Harms Living Things

Ozone molecules wind up near the Earth's surface as a part of air pollution. Ozone molecules near the ground damages lung tissues of animals and prevent plant respiration by blocking the openings in leaves where respiration occurs. Without respiration, a plant is not able to photosynthesize at a high rate and so it will not be able to grow.

Global Warming Harms Living Things

Our planet is currently warming much more rapidly than expected because additional greenhouse gasses are being released into the atmosphere from air pollution. When fuels are burned, some of the pollutants released are greenhouses gasses. Through the process of photosynthesis, plants convert carbon dioxide into oxygen and use the carbon to grow larger. However, the amount of carbon dioxide released by burning fuels is much more than plants can convert.

Global warming is causing changes to the places where plants and animals live around the world. For example:

- Near the poles, ice and frozen ground are melting. This causes changes in the habitat and resources for plants and animals living there.
- Ocean warming, rising sea levels, runoff, and coral diseases are causing change in shallow marine environments such as coral reefs.
- Global warming is causing less rain to fall in the middle of

continents. This makes these areas very dry and limits water resources for plants and animals.

AIR POLLUTION AND WATER

Have we heard about rivers, lakes, or streams becoming polluted? Sometimes the pollution comes from trash, oil spills, sewage, fertilizers, or chemicals. However, sometimes the source of water pollution is in the air.

Air pollution can make its way into rivers, lakes, or streams. Several different types of air pollutants are able to waterways. Some fall from the sky as dry particles. Other air pollutants are carried to the ground in raindrops, snowflakes, or fog. Some of these pollutants are listed below. They not only harm water, but also the plant and animal life that depend on water to survive.

Nitrogen Compounds

Nitrogen is a nutrient that plants need to grow. However, there can be too much of a good thing. Too much nitrogen in a body of water can cause algae to grow very quickly, clogging the waterways, and upsetting the balance of the ecosystem. This is called an algal bloom or "Red Tide." Algal blooms happen more often now than they did hundreds of years ago. Some algal blooms are toxic. When animals eat the algae they also eat the toxins. Nitrogen compounds in air pollution are partly the cause of algal blooms, and can also contribute to water bodies becoming more acidic.

Sulfur Dioxide and Nitrogen Oxides

When fossil fuels are burned, sulfur dioxide and nitrogen oxides are released into the atmosphere. Both of these air pollutants dissolve in water vapour to form acid. The acidic water vapour condenses into clouds and falls eventually as precipitation such as rain or snow. This is known as "acid rain".

It falls to Earth and eventually enters bodies of water making them more acidic. That is a tough environment for some fish and animals, such as frogs, to survive and reproduce within. Acidic waters prevent fish eggs from hatching. In fact, some very acid lakes have no fish at all.

Some types of rocks neutralize acidic water. The acidic water dissolves rocks such as limestone. The limestone is then in the water, and makes the water less acidic. Bodies of water in areas such as the Northeastern United States and Eastern Canada are in more danger because rock is primarily granite, which does not neutralize acid rain.

Mercury

Mercury occurs naturally in the environment but it is also released into the atmosphere by people, mainly from burning waste, and especially from burning fossil fuels such as coal. Mercury can dissolve in water, where bacteria in the water transform it into poisonous methyl mercury. Fish and shellfish absorb methyl mercury into their bodies. When other animals, such as birds or people, eat the fish, the methyl mercury gets into their bodies as well. The United States and other countries test to ensure that the fish sold in stores are safe to eat.

How can people reduce air pollution's effect on water? Reducing our use of fossil fuels can make the biggest impact. Turn off lights. Walk, ride a bike, or use public transportation. Every little bit done by each person can add up to a noticeable improvement in air pollution.

Chapter 14

Impact of Ozone Layer

The ozone molecule consists of three oxygen atoms that are bound together (triatomic oxygen, or O_3). Unlike the form of oxygen that is a major constituent of air (diatomic oxygen, or O_2), ozone is a powerful oxidizing agent. Ozone reacts with some gases, such as nitric oxide or NO, and with some surfaces, such as dust particles, leaves, and biological membranes. These reactions can damage living cells, such as those present in the linings of the human lungs. Exposure has been associated with several adverse health effects, such as aggravation of asthma and decreased lung function.

Ozone was first observed in the Los Angeles area in the 1940s. The ozone that the ARB regulates as an air pollutant is mainly produced close to ground (tropospheric ozone), where people live, exercise, and breathe. A layer of ozone high up in the atmosphere, called stratospheric ozone, reduces the amount of ultraviolet light entering the earth's atmosphere. Without the protection of the stratospheric ozone layer, plant and animal life would be seriously harmed. In this document, 'ozone' refers to tropospheric ozone unless otherwise specified.

Most of the ozone in California's air results from reactions between substances emitted from vehicles, industrial plants, consumer products, and vegetation. These reactions involve volatile organic compounds (VOCs, which the ARB also refers to as reactive organic gases or ROG) and oxides of nitrogen (NOx) in the presence of sunlight. As a photochemical pollutant, ozone is formed only during daylight hours under appropriate conditions, but is destroyed throughout the day and night. Therefore, ozone concentrations vary depending upon both the time of day and the location. Ozone concentrations are higher on hot, sunny, calm days. In metropolitan areas of California, ozone concentrations frequently exceed regulatory standards during the summer.

From the 1950s into the 1970s, California had the highest ozone concentrations in the world, with hourly average concentrations in Los

Angeles peaking over 0.5 ppm and frequent "smog alerts". In the early 1970s, the ARB initiated emission control strategies that provided for concurrent and continuing reductions of both NOx and VOC from mobile sources and, in conjunction with the local air districts, stationary and area sources. Since then, peak ozone concentrations have decreased by more than 60 per cent and smog alerts no longer occur in the Los Angeles area, despite more than a 35 per cent increase in population and almost a doubling in vehicle miles traveled. However, most Californians still live in areas that do not attain the State's health-based standard (0.09 ppm for one hour) for ozone in ambient air.

Formation and Removal of Tropospheric Ozone

The formation of ozone in the troposphere is a complex process involving the reactions of hundreds of precursors.

Nitrogen Cycle and the Photostationary-State Relationship for Ozone

The formation of ozone in the troposphere results from only one known reaction: addition of atomic oxygen (O) to molecular oxygen (O_2) in the presence of a third "body" (M). [M is any "body" with mass, primarily nitrogen or oxygen molecules, but also particles, trace gas molecules, and surfaces of large objects. M absorbs energy from the reaction as heat; without this absorption, the combining of O and O_2 into O_3 cannot be completed.]

$$O + O_2 + M \rightarrow O_3 + M$$

The oxygen atoms are produced primarily from photolysis of NO_2 by the ultraviolet portion of solar radiation (hn).

$$NO_2 + hn \rightarrow NO + O$$

Reaction 3 converts ozone back to oxygen and NO back to NO_2, completing the "nitrogen cycle."

$$O_3 + NO \rightarrow NO_2 + O_2$$

Reactions 1 and 3 are comparatively fast. Therefore, the slower photolysis reaction 2 is usually the rate-limiting reaction for the nitrogen cycle and the reason why ozone is not formed appreciably at night. It is also one of the reasons why ozone concentrations are high during the summer months, when temperatures are high and solar radiation is intense. The cycle time for the three reactions described above is only a few minutes. Ozone accumulates over several hours, depending on emission rates and meteorological conditions. Therefore, the nitrogen cycle operates fast enough to maintain a close approximation to the following photostationary-state equation derived from the above reactions.

$[O_3]$photostationary-state = (k2/k3) × $[NO_2]/[NO]$ (the brackets denote concentration)

The ratio of the rate constants for reactions (k2/k3), is about 1:100. Assuming equilibrium could be reached in the ambient air and assuming typical urban pollution concentrations, a NO_2 to NO ratio of 10:1 would be needed to generate about 0.1 ppm of ozone (a violation of the state one-hour ozone standard [0.09 ppm]). In contrast, the NO_2 to NO emission ratio is approximately 1:10; therefore, the nitrogen cycle by itself does not generate the high ozone concentrations observed in urban areas. The net effect of the nitrogen cycle is neither to generate nor destroy ozone molecules. Therefore, for ozone to accumulate according to the photostationary-state equation, an additional pathway is needed to convert NO to NO_2; one that will not destroy ozone. The photochemical oxidation of VOCs, such as hydrocarbons and aldehydes, provides that pathway.

The VOC Oxidation Cycle

Hydrocarbons and other VOCs are oxidized in the atmosphere by a series of reactions to form carbon monoxide (CO), carbon dioxide (CO_2) and water (H_2O). Intermediate steps in this overall oxidation process typically involve cyclic stages driven by hydroxyl radical (OH) attack on the parent hydrocarbon, on partially oxidized intermediate compounds, and on other VOCs. The Hydroxyl radical is ever-present in the ambient air; it is formed by photolysis from ozone in the presence of water vapour, and also from nitrous acid, hydrogen peroxide, and other sources. In the sequence shown below, R can be hydrogen or virtually any organic fragment. The oxidation process usually starts with reaction, from OH attack on a hydrocarbon or other VOC:

$$RH + OH \text{ à } H_2O + R$$

This is followed by reaction with oxygen in the air to generate the peroxy radical (RO_2).

$$R + O_2 + M \text{ à } R\dot{O}_2 + M$$

The key reaction in the VOC oxidation cycle is the conversion of NO to NO_2. This takes place through the fast radical transfer reaction with NO.

$$RO_2 + NO \text{ à } NO_2 + RO$$

R can also be generated by photolysis, which usually involves only VOCs with molecules containing the carbonyl (C=O) bond. The simplest VOC molecule that contains the carbonyl bond is formaldehyde (HCHO). Because formaldehyde enters into several types of reactions of importance for understanding ozone formation and removal, we will use it to help illustrate these reactions. The oxidation cycle for formaldehyde can be written in the following sequence of reactions.

$OH + HCHO \rightarrow H_2O + HCO$

$HCO + O_2 \rightarrow HO_2 + CO$

$HO_2 + NO \rightarrow NO_2 + OH$

Hydroperoxyl radical (HO_2) is generated by reaction, and the hydroxyl radical returns in reactio to complete the cycle. In addition, reaction produces the NO_2 required for ozone formation, as described above. Also, the carbon monoxide (CO) generated by reaction 8 can react like an organic molecule to yield another hydroperoxyl radical.

$OH + CO \rightarrow H + CO_2$

$H + O_2 + M \rightarrow HO_2 + M$

Another component that formaldehyde provides for smog formation is a source of hydrogen radicals.

HCHO + hn→ H + HCO

The hydrogen atom (H) and formyl radical (HCO) produced by this photolysis reaction yield two hydroperoxyl radicals via reaction with oxyge.

The reactions above comprise the simplest VOC oxidation cycle. Actually, hundreds of VOC species participate in thousands of similar reactions.

The Nitrogen Dioxide and Radical Sink Reaction

Another reaction is central to a basic understanding of ozone formation: the NO_2 plus radical sink reaction that forms nitric acid.

NO_2 + OH + M à HNO_3 + M

The previous discussion can be used to explain the typical pattern of ozone concentrations found in the urban atmosphere. Nitric oxide concentrations are relatively high in the early morning because the free radicals needed to convert the NOx emissions (which are primarily NO) to NO_2 are not yet present in sufficient quantities.

After sunrise, photolysis of formaldehyde (reaction 12) and other compounds starts the VOC oxidation cycle for the hundreds of organic gases present in the atmosphere. Subsequent NO to NO_2 conversion by the peroxy radical (reaction 6) results in NO_2 becoming the dominant NOx species.

When the NO_2 to NO ratio becomes large enough, ozone builds up. In the South Coast Air Basin (Los Angeles area), the highest ozone concentrations are observed in the San Bernardino Mountains, many miles downwind from the highest concentration of emission sources (freeways, power generating facilities, and oil refineries along the coast), because the reactions involving the organic gases are relatively slow. Meanwhile, NO_2 concentrations decrease via the sink reaction 13.

Winds disperse and dilute both NOx and ozone. During the day, NOx is also diluted by the diurnal rising of the inversion layer, allowing for more mixing. For ozone, however, the deepening mixing layer may cause its concentration to decrease on some days and increase on others. Although increased mixing almost always dilutes NOx, the effect of increased mixing on ozone concentrations depends upon whether higher concentrations of ozone are present aloft. Ozone that is trapped above the inversion layer overnight is available to increase the concentrations of ozone generated by the following day's emissions.

During the night, NO and ozone combine to form NO_2 and oxygen via reaction 3 until either the NO or ozone is consumed. Nitrous acid or HONO is also present at night in polluted ambient air in California. Nitrous acid is produced from NO_2 and water, and is also emitted from various combustion sources. Its levels are low during the day because sunlight breaks it down rapidly. At sunrise, sunlight causes gas-phase HONO to react rapidly to provide NO and OH, two key reactants in the formation of ozone. In this way, they help initiate ozone formation in the morning by being available to react with VOCs as soon as their emissions increase due to an increase in human activity.

Nitric acid (HNO_3) was once thought to be a permanent sink for NOx and for radicals. However, nitric acid on surfaces may react with NO to regenerate NO_2, which would increase the ozone-forming potential of NOx emissions.

Ratio of Volatile Organic Compounds

Although VOCs are necessary to generate high concentrations of ozone, NOx emissions can be the determining factor in the peak ozone concentrations observed in many locations. VOCs are emitted from both natural and anthropogenic sources. Statewide, natural VOC sources dominate, primarily from vegetation. However, in urban and suburban areas, anthropogenic VOC emissions dominate and, in conjunction with anthropogenic NOx emissions, lead to the peak concentrations of ozone observed in urban areas and areas downwind of major urban areas. The relative balance of VOCs and NOx at a particular location helps to determine whether the NOx behaves as a net ozone generator or a net ozone inhibitor. When the VOC/ NOx ratio in the ambient air is low (NOx is plentiful relative to VOC), NOx tends to inhibit ozone formation. In such cases, the amount of VOCs tends to limit the amount of ozone formed, and the ozone formation is called "VOC-limited". When the VOC/ NOx ratio is high (VOC is plentiful relative to NOx), NOx tends to generate ozone. In such cases, the amount of NOx tends to limit the amount of ozone formed, and ozone formation is called "NOx -limited". The VOC/ NOx ratio can differ substantially by location and time-of-day within a geographic area. Furthermore, the VOC/ NOx ratio measured near the ground might not represent the ratio that prevails in the air above the ground where most of the tropospheric ozone is generated.

Photochemical Reactivity

Photochemical reactivity, or reactivity, is a term used in the context of air quality management to describe a VOC's ability to react (participate in photochemical reactions) to form ozone in the atmosphere. Different VOCs react at different rates. The more reactive a VOC, the greater potential it has to form ozone. Examples of the more reactive VOCs in California's atmosphere include propene, *m*-xylene, ethene, and formaldehyde. The ARB has helped to pioneer an approach to ozone control that considers the reactivity of each VOC constituent. In California's urban areas, ozone formation tends to be limited by the availability of VOCs. Therefore, the reactivity-based regulatory approach has been applied in conjunction with reduction of NOx emissions. Reactivity-based regulations promote the control of those VOCs that form ozone most effectively, thereby guiding the affected industries (such as manufacturers of motor vehicle and consumer product formulators that use solvents) to choose the most cost-effective processes and designs to reduce VOC emissions.

Role of Weather in Ozone Air Quality

In the troposphere, the air is usually warmest near the ground. Warm air has a tendency to rise and cold air to sink, causing the air to mix, which disperses ground-level pollutants. However, if cooler air gets layered beneath warm air, no mixing occurs — the air is stable or stagnant. The region in which temperature is so inverted is called an inversion layer. One type of inversion occurs frequently several thousand feet above the ground and limits the vertical dispersion of pollutants during the daytime. Another type of inversion occurs on most evenings very near the ground and limits the vertical dispersion of pollutants to a few hundred feet during the night. Pollutants released within an inversion tend to get trapped there. When the top of the daytime inversion is especially low [in elevation], people can be exposed to high ozone concentrations. Mountain chains, such as those downwind of California's coastal cities and the Central Valley, help to trap air and enhance the air quality impact of inversions. Cooler air draining into the state's valleys and 'air basins' also enhances inversion formation.

The direction and strength of the wind also affect ozone concentrations. Based on worldwide climate patterns, western coasts at California's latitude tend to have high-pressure areas over them, especially in summer. By preventing the formation of storms, and by promoting the sinking of very warm air, these high-pressure areas are associated with light winds and temperature inversions, both of which limit dispersion of pollutants.

Because tropospheric ozone forms as a result of reactions involving other pollutants, the highest concentrations tend to occur in the afternoon. The photochemical reactions that create ozone generally require a few hours after the emissions of substantial VOC emissions, and are most effective when sunlight is intense and air temperatures are warm.

Ozone concentrations in California are usually highest in the summer. The prevailing daytime winds in summer are on-shore, bringing relatively clean air from over the ocean to the immediate coastal areas, but carrying emissions of ozone precursors further inland. With the climatically favored clear skies and temperature inversions that limit the vertical dispersion of pollutants, these emissions are converted into ozone, with the highest concentrations tending to occur at distances a few tens of miles downwind of urban centers (ARB 2002).

During the periods of the year when the sunlight is most intense, much of California experiences a high frequency of inversions, relatively low inversion heights, and low wind and rainfall. As a result, no other State has more days per year with such a high potential for unhealthy ozone concentrations.

Spatial and Temporal Variations of Ozone Concentrations

Spatial Variations of Ozone Concentrations

Ambient ozone concentrations can vary from non-detectable near combustion sources, where nitric oxide (NO) is emitted into the air, to several hundreds parts per billion (ppb) of air in areas downwind of VOC and NOx emissions. In continental areas far removed from direct anthropogenic effects, ozone concentrations are generally 20 - 40 ppb. In rural areas downwind of urban centers, ozone concentrations are higher, typically 50 - 80 ppb, but occasionally 100 - 200 ppb. In urban and suburban areas, ozone concentrations can be high (well over 100 ppb), but peak for at most a few hours before deposition and reaction with NO emissions cause ozone concentrations to decline (Finlayson-

Pitts and Pitts 2000, Seinfeld and Pandis 1998, Chameides et al. 1992, Smith et al. 1997).

Ozone concentrations vary in complex ways due to its photochemical formation, its rapid destruction by NO, and the effects of differing VOC/ NOx ratios in air. A high ratio of NOx emissions to VOC emissions usually causes peak ozone concentrations to be higher and minimum concentrations to be lower, compared to background conditions. Peak ozone concentrations are usually highest downwind from urban centers. Light winds carry ozone from urban centers, and photochemical reactions create ozone from urban emissions of VOC and NOx. Also, away from sources of NOx emissions, less NO is available to destroy ozone. Due to the time needed for transport, these peak ozone concentrations in downwind areas tend to occur later in the day compared to peak ozone concentrations in urban areas.

Due to the lack of ozone-destroying NO, ozone in rural areas tends to persist at night, rather than declining to the low concentrations (<30 ppb) typical in urban areas and areas downwind of major urban areas, that have plenty of fresh NO emissions. Ratios of peak ozone to average ozone concentrations are typically highest in urban areas and lowest in remote areas. Within the ground-based inversions that usually persist through the night, ozone concentrations can be very low. In urban areas, emissions of NO near the ground commonly reduce ozone below 30 ppb. In rural areas, however, NO emissions are less prevalent and nighttime ozone may persist well above 30 ppb.

Temporal Variations in Ozone Concentrations

Ambient ozone concentrations tend to vary temporally in phase with human activity patterns, magnifying the resulting adverse health and welfare effects. Ambient ozone concentrations increase during the day when formation rates exceed destruction rates, and decline at night when formation processes are inactive. This diurnal variation in ozone depends on location, with the peaks being very high for relatively brief periods of time (an hour or two duration) in urban areas, and being low with relatively little diurnal variation in remote regions. In urban areas, peak ozone concentrations typically occur in the early afternoon, shortly after solar noon when the sun's rays are most intense, but persist into the later afternoon, particularly where transport is involved. Thus, the peak urban ozone period of the day can correspond with the time of day when people, especially children, tend to be active outdoors.

In addition to varying during the day, ozone concentrations vary during the week. In the 1960s, the highest ozone concentrations at many urban monitoring sites tended to occur on Thursdays. This pattern was believed to be due to the carryover of ozone and ozone precursors from one day to the next, resulting in an accumulation of ozone during the workweek. In the 1980s, the highest ozone concentrations at many sites tended to occur on Saturdays and the "ozone weekend effect" became a topic of discussion. Since then, the weekend effect has become prevalent at more urban monitoring locations and the peak ozone day of the week has shifted to Sunday. Although ozone concentrations have declined on all days of the week in response to emission controls, they have declined faster on weekdays than on weekends. Thus, the peak ozone period of the week now tends to coincide with the weekend, when more people tend to be outdoors and active than during the week.

The causes of the ozone weekend effect and its implications regarding ozone control strategies have not yet been resolved. Almost all of the available data represent conditions at ground level, where the destruction of ozone by fresh emissions of NO is a major factor controlling ozone concentration. However, most ozone is formed aloft, and the air quality models used to analyse ozone formation have not demonstrated the ability to represent the ozone-forming system aloft with sufficient realism. In addition, several potentially significant photochemical processes are yet to be fully incorporated in simulation models. These deficiencies leave unrèsolved this fundamental question: does the ozone weekend effect occur because more ozone is formed (aloft) on weekend, because more ozone is destroyed (at the surface) on weekdays, or because ozone formation is more efficient on weekends.

Ozone concentrations also vary seasonally. Ozone concentrations tend to be highest during the summer and early fall months. In areas where the coastal marine layer (cool, moist air) is prevalent during summer, the peak ozone season tends to be in the early fall. Additionally, as air pollution controls have reduced the emissions of ozone precursors and the reactivity of VOCs, ozone concentrations have declined faster during times of the year when temperatures and the amount of sunlight are less than during the summer. Thus, the peak ozone season corresponds with the period of the year when people tend to be most active outdoors.

Also, ozone concentrations can vary from year to year in response to meteorological conditions such as El Niño and other variations in global pressure systems that promote more or less dispersion of emissions than typical. Although peak ozone concentrations vary on a year-to-year basis, peak ozone concentrations in southern California have been declining on a long-term basis, as anthropogenic emissions of VOC and NOx have declined. However, since the advent of the industrial revolution, global background concentrations of ozone appear to be increasing. This increase has implications regarding the oxidative capability of the atmosphere and potentially global warming processes (ozone is a strong greenhouse gas but is present at relatively low concentrations).

STRATOSPHERIC AEROSOL

Aerosol in the stratosphere can be seen with the naked eye (in the form of luminous sunsets following large volcanic eruptions) only a few times over the course of a lifetime. Similarly, consider that under nominal non-volcanic background conditions that the stratosphere contains about 1 Tg (1 megatonne). If this material were deposited uniformly onto the surface of the Earth, it would result in a layer only about 1-nm thick or less than one ten-thousandth of the width of a human hair. With this in mind, it is not difficult to image that the general public may not appreciate the important role that stratospheric aerosol can play in climate. However, in this era of shrinking science dollars, it is required to develop coherent arguments for continued research and investment into what is almost by definition an esoteric field.

The Source of Stratospheric Aerosol

Since routine measurement of stratospheric aerosol began in the late 1970's, aerosol has primarily originated from episodic injections of SO_2, a gaseous precursor to sulfate aerosol. Volcanic eruptions can inject a large amount of this gas into the stratosphere. For instance, the 1991 eruption of Mt. Pinatubo in the Philippines increase the total stratospheric aerosol mass by a factor of approximately 30 to about 30 Tg. Other noteworthy eruptions include Tambora (Indonesia) in 1815, Krakatau (Indonesia) in 1883, and El Chichon (Mexico) in 1982. Not all eruptions result in a significant impact on stratospheric aerosol levels. For instance, Mt. St. Helens (US) in 1980 and Nevado del Ruiz (Colombia) in 1985, which killed over 25000 people, had only modest effects on the stratosphere.

The non-volcanic sources of aerosol include the transport of carbonyl sulfide (OCS), a product of biologic activity, from the lower atmosphere in the stratosphere where it is photochemically transformed into sulfate aerosol. Most estimates of the strength of this source indicate that it is insufficient to explain aerosol levels observed even in the cleanest periods of the last 20 years. As a result, alternative explanations have been suggested such as human-derived aerosol from industry and/or aircraft. Or that the time constant for stratospheric aerosol removal following an eruption is sufficiently long that, given the magnitude of the volcanic perturbations, that a non-volcanic background has simply not been attained during this period. In fact, an outstanding scientific question is whether human activities significantly impact aerosol levels in the lower stratosphere. Projections for future ozone trends and the future of stratospheric air travel including projected supersonic fleets will be strongly impacted by the results of research.

Effect on Climate

The effect of stratospheric aerosol on climate is at its simplest level straightforward to understand. Under "background" conditions, stratospheric aerosol does not play a significant role in the radiative processes of the Earth-atmosphere system.

The presence of enhanced aerosol following an eruption reflects in-coming solar radiation back into space and, as a result, acts to cool the Earth. In the details, it is rather more complicated. Since the aerosol absorbs in the infrared, upwelling radiation from the Earth's surface is absorbed in the stratosphere, which is warmed as a result. In fact, this is perhaps the most readily discerned effect immediately after an eruption as temperatures rise at some levels well beyond climatological ranges.

For instance, following the 1991 eruption of Mt. Pinatubo temperatures measured by radiosonde from Singapore were more than x K greater than the expected value at altitudes near 30 km (or more 3 standard deviations). Similarly, using either ground-based or satellite-based measurements, it can be demonstrated that global-average surface temperature decreased following the Pinatubo eruption. However, the decrease was not uniformly distributed and with some areas significantly cooler than average others apparently unaffected or even warmer than average in the months following the eruption.

Both observations and modeling seem to indicate that there are regions that are more susceptible to cooling following large eruptions. The northeastern US and Western Europe are particularly effected such that the well-known "year without a summer" followed the Tambora eruption of 1815. In fact, the weather in northern Europe that year was so uniformly dreary that it may have contributed to Shelley's authorship of Frankenstein. In addition, the gradual warming during the late 1980's and early 1990's thought to be the result of green houses was interrupted by the 1991 Pinatubo eruption.

Effect on Ozone Chemistry

Aerosol in the stratosphere can have a significant effect on the chemical processes that govern ozone concentration. Classic, gas-phase only chemistry predicts that ozone depletion related to chlorofluorocarbons (CFCs) predicts that the greatest loss of ozone should occur in the middle-stratosphere near 30 km. In fact, the greatest observed relative trends in ozone occur in the lower stratosphere where chemical processes that occur on the surface of aerosol (called heterogeneous processes) significantly modify the reactive chlorine balance and enhance ozone destruction. Recent research demonstrates that even small enhancements over "background" aerosol levels are sufficient to repartition chlorine from relatively non-reactive (e.g., $ClONO_2$) forms to ones that ones detrimental to the ozone balance (e.g., ClO). In fact, it is a reasonable assertion to state that it is impossible to correctly assess long-term ozone trends apart from a concomitant understanding of aerosol variability.

Chapter 15

Reduction of Greenhouse Gases

THE FOOD SYSTEM

The food system encompasses all the activities associated with providing food to consumers. These include food production (e.g., onsite clearing, cultivation, and harvesting, and offsite production of fertilizer and other agricultural inputs) and post-harvest activities (e.g., food processing, transport, cooking).

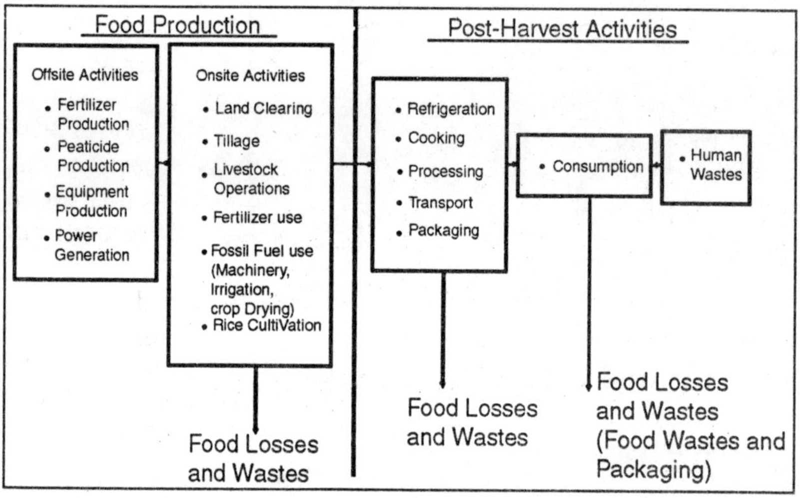

Fig. The Food System

Congress has become increasingly aware of the environmental impacts of the food system and has begun to address some of these through legislation. In addition to previously identified environmental impacts of agriculture (pollution of surface and groundwater by nitrate and pesticides, soil erosion, depletion of aquifers to meet irrigation needs, loss of natural habitat), the food system is now also recognized as a potential contributor to global climate change.

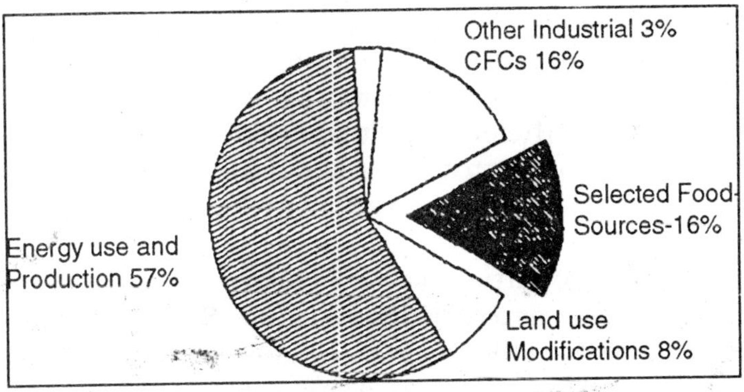

Fig. Contribution of Selected Food-Related

Figure shows the global food system's contribution to global warming in the 1980s. Although estimates are uncertain, the food sector may account for one-third of global methane (CH_4) emissions; one-fifth of net global carbon dioxide (CO_2) emissions; up to 15 per cent of chlorofluorocarbon (CFC) emissions; and anywhere from one-tenth to one-fifth of current global nitrous oxide (N_2O) emissions. Food sector emissions of all these gases will grow, barring efforts to contain them, as efforts to provide food for the world's growing population intensify.

Uncertainty in agricultural emissions data currently makes it difficult to predict the efficacy of any of the control methods available to reduce this sector's contribution to global warming, yet many of these controls deserve consideration in their own right as a means to combat other agriculture-related environmental problems. Indeed, many have been or are being considered by Congress for reasons other than climate change.

In the United States and the industrialized world in general. several options are available to reduce food sector emissions in the near term. Methane emissions from livestock could be reduced by improving nutrient and manure management (and, possibly, by increasing productivity) or by reducing demand for livestock products. Fertilizers and other sources of applied nitrogen, such as crop residues and animal wastes, could be used more efficiently; this may reduce N_2O emissions, as well as surface and groundwater contamination, and would help conserve soil organic matter. Nitrogen fertilizer manufacturing and onsite farm machinery and cultivation practices could be more energy efficient; while reductions in CO_2 emissions would be relatively small, other benefits such as decreased local air pollution from fossil fuel combustion would accrue. Land transformations that help remove carbon from the atmosphere (such as converting cropland to forest land) could be encouraged, while those that increase CO_2 emissions could be discouraged. In food refrigeration, emissions can be curbed by preventing the release of CFCs from existing refrigerators and eliminating their future use; and by improving energy efficiencies. Further CO_2 reductions could be achieved by designing stoves and ovens that use energy more efficiently and by increasing fuel efficiency in vehicles used in food transport.

In developing countries, slowing deforestation, maintaining or increasing crop yields, and reducing emissions associated with cooking can be more effective, in the short term, than changing current patterns of fossil fuel use. Alternatives to clearing of tropical forests include increased use of agroforestry and "sustainable" agriculture, and decreased subsidies for cattle ranching on lands that cannot support livestock for more than a few years.

Crop yields may be improved on existing agricultural land with increased agricultural inputs (e.g., fertilizers, irrigation, etc.), but this may increase greenhouse gas emissions per acre. Opportunities exist to reduce methane emissions from livestock through technology transfer to developing countries, but the relevant technologies may not be readily applicable in most developing countries. In general, direct U.S. Government influence in these areas tends to be through the U.S. Agency for International Development (A.I.D.), research programs, multinational corporations, and participation in multilateral lending institutions.

EMISSIONS FROM THE FOOD SYSTEM

Activities in the food system affect the flows of many substances to and from the atmosphere, including small particles (aerosols) and numerous trace gases.

The system itself and key trends in global food production and consumption. In some developing countries, food production (i.e., activities up to and including harvest) is the dominant source of greenhouse gases, primarily because of CO_2 emitted through land transformations (e.g., land clearing and field burning) and CH_4 emitted through rice cultivation. Although global data on emissions from post-harvest activities are poor, these activities are likely the most important source of emissions in *industrialized countries.*

Food Production

Greenhouse gas sources from food production activities include:

- Flooded rice fields, which are significant sources of CH_4, particularly in the developing countries of Asia;

- Livestock, a significant source of CH_4 in many industrialized and developing countries, through direct emissions from animals as well as their manure;
- Nitrogenous fertilizers, the use of which results in N_2O;
- Large-scale land transformations (e.g., clearing tropical forests) in many developing countries; and, to a lesser extent, land use changes in industrialized countries (e.g., urbanization), both of which result in CO_2 emissions; and
- Burning of vegetation to clear and/or prepare land for agriculture (especially in developing countries), which adds to atmospheric concentrations of several gases, including CO_2, CH_4, N_2O, and carbon monoxide.

In general, emissions from food production and their contributions to climate change are difficult to quantify because of the complex interactions of biological and chemical processes in soils, water, plants, animals, and the atmosphere, and because studies are lacking for many topics and areas.

Nevertheless, some approximations and projections are possible. For example, EPA estimates that if current agricultural practices continue, CH_4 emissions from rice will increase by about 35 per cent by 2025, those from livestock will rise by about 65 per cent, and N_2O emissions from fertilizers could more than double.

Methane (CH_4)

Methane is produced when bacteria decompose organic material in oxygen-deficient (i.e., anaerobic) environments, such as sediments at the bottom of flooded or rainfed rice paddies, landfills, and the digestive tract of ruminant animals and termites; CH_4 is also emitted as a byproduct when wood and other biomass are burned.

Emissions from these and other CH_4 sources are very poorly characterized, partly because they vary enormously on geographic scales and over time.

Available evidence indicates that about one-third of total global CH_4 emissions comes from the food sector; this represents about 60 per cent of total emissions from anthropogenic sources, or roughly 10 per cent of global warming in the 1980s.

Rice Paddies

Rice provided about one-fourth of the world's cereal grains in 1985. China and India produced well over half of the total, the United States only 1 per cent. Global rice production increased by nearly 200 per cent between the Late 1940s and 1985, while acreage grew by 43 per cent.

Estimates of CH_4 released from rice paddies in the mid-1980s range from roughly 25 to 170 million metric tons, or about 15 to 30 per cent of world methane emissions from anthropogenic sources.

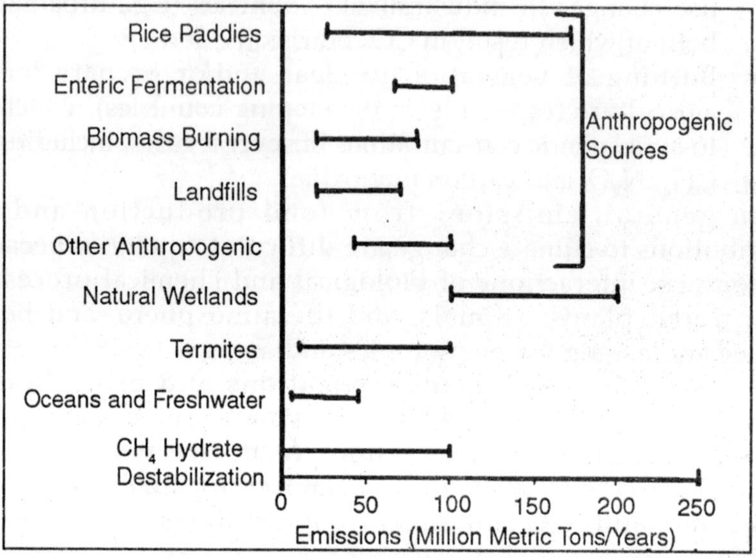

Fig. Estimated Global Emissions of Methane

Rice paddies thus accounted for about 3 to 6 per cent of the contribution to global warming in the 1980s. These emissions estimates are based on a few studies conducted in temperate regions (California, Italy, and Spain), but because CH_4 emissions are likely to be highly site-specific, the extrapolation of these data to other regions (including tropical areas, where most of the world's rice is grown) is problematic. However, new data from Japan, China, and the Philippines are now becoming available.

Recent field tests, for example, have shown that during the growing season, rice fields in China emit up to 10 times more CH_4 per hectare per hour than rice fields in Europe and the United States. If new data from China are representative of conditions in the Far East (over 90 per cent of the world's rice is produced in Asia), then CH_4 emissions from rice cultivation are higher than the estimates presented above, although it is difficult to estimate by how much.

Ruminants

Much of the world's livestock consists of ruminant animals—

sheep, goats, camels, cattle, and buffalo. One of the unique features of ruminants is their four-chambered stomach, including one chamber called the rumen in which bacteria break down food and generate CH_4 as a byproduct. Ruminants emitted an estimated 65 to 100 million metric tons of CH_4 per year in the mid-1980s, perhaps 10 to 20 per cent of global CH_4 emissions from all sources, or about 20 to 40 per cent of total anthropogenic emissions. Therefore, ruminant digestion accounted for about 4 to 7 per cent of the global warming in the 1980s, given that all anthropogenic CH_4 sources accounted for a total of about 19 per cent. Globally, cattle account for about three-fourths of livestock CH_4 emissions, or about 7 to 15 per cent of total global CH_4 emissions from all sources. Beef and dairy cattle in the United States account for about 1 per cent of total global CH_4 emissions.

The above estimates do not include CH_4 emissions from animal manure. If manure decomposes anaerobically, some of its organic matter is converted to CH_4. In industrialized countries, manure handling practices at feedlots, dairies, and swine and poultry farms may release significant methane. In developing countries, however, most manure is spread as fertilizer, burned for fuel, or left in pastures; the magnitude of CH_4 emissions from these practices is poorly quantified but likely is low since most decomposition takes place aerobically (however, this leads to more CO_2 emissions). Preliminary estimates suggest global CH_4 emissions from manure are on the order of about 20 to 40 million metric tons per year.

Biomass Burning

Burning vegetation contributes 20 to 80 million metric tons of methane per year, or roughly 7 to 8 per cent of global emissions from anthropogenic sources and natural sources such as lightning-induced forest fires. This portion may be higher if recent data indicating higher deforestation rates are correct. Substantial but unquantified CH_4 emissions result from fires ignited deliberately—forests burned to produce rangeland or cropland; grasslands burned to enhance forage; and crop residues burned to return nutrients to the soil. These emissions are heavily concentrated in tropical countries, where large areas of savanna and forest are burned or cleared each year for agriculture.

NITROUS OXIDE (N_2O)

Agriculture introduces nitrogenous compounds to the environment in the form of commercial fertilizers, legumes, and crop residues. N_2O emissions from soil and water occur through nitrification and denitrification of these compounds and also result when vegetation is cleared through burning.

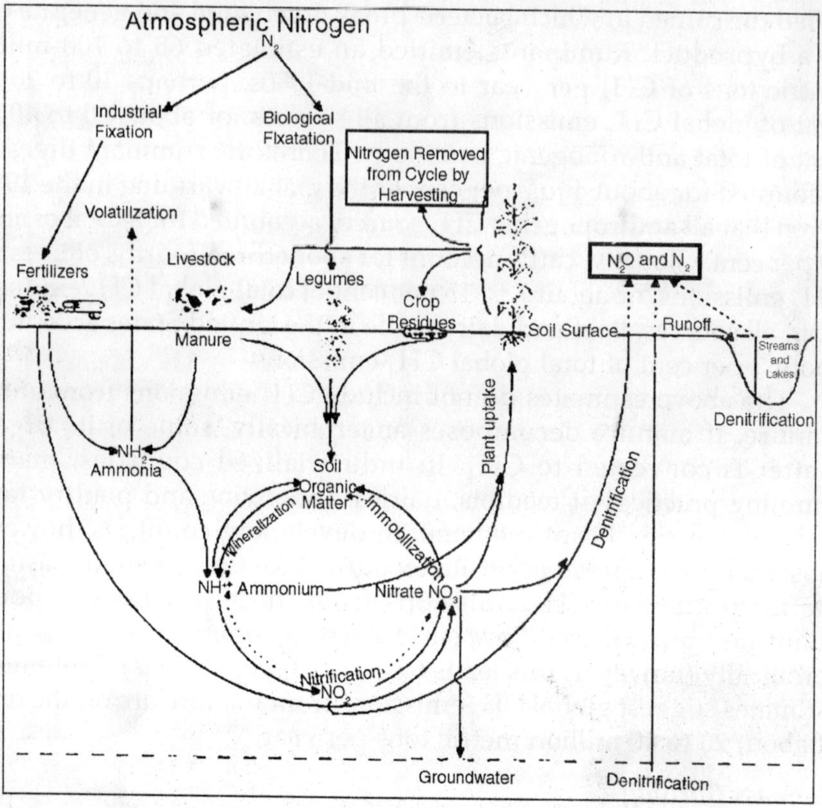

Fig. Nitrogen Cycle

The magnitude of N_2O emissions from terrestrial and aquatic sources is very poorly characterized. Based on annual increases in N_2O atmospheric concentrations, the Intergovernmental Panel on Climate Change estimates that total global N_2O emissions *should* be around 10 to 17.5 million metric tons per year; however, only 4.4 to 10 million metric tons can be accounted for from known sources.

N_2O emissions associated with fertilizer use are not well understood and probably vary with factors such as fertilizer type, method of application, and amount applied. Worldwide nitrogen fertilizer consumption is about 74 million metric tons per year. China accounted for nearly 20 per cent of this consumption in 1987, the United States for about 13 per cent. Current fertilizer-derived emissions are on the order of 0.01 to 2.2 million metric tons per year, about 0.2 to 20 per cent of global emissions from all sources, or about 10 to 80 per cent of all anthropogenic emissions.

The general relationship between nitrogen application rate and maize yields illustrates that yields are highest at a certain optimal nitrogen application rate; further additions result in either stable or even lower yields (because nitrogen is no longer used as effectively by plants), and nitrogen concentrations build up in the soil. The potential for nitrogen losses through leaching, volatilization, or denitrification grows.

Nitrogen application rates are generally higher in developed countries. Although U.S. application rates for wheat are comparatively lower than those for many countries, U.S. rates for other crops (e.g., corn, rice) are among the highest in the world. Although global fertilizer application data provide a picture of overall intensity of fertilizer use, they do not reveal whether nitrogen fertilizers generally are being over- or under-applied for specific crops and countries.

Leguminous crops (e.g., soybeans, peas, alfalfa) also add nitrogen to agricultural soils; legumes use atmospheric nitrogen directly and require much less nitrogenous fertilizers than non-leguminous crops. Worldwide production of legumes increased by roughly 85 per cent from the late 1940s to 1985. Two-thirds of the world's legume production is in developing countries; only 2 per cent is in the United States.

CARBON DIOXIDE (CO_2)

The flow of CO_2 to and from the atmosphere is influenced by food production in two ways:

- Changes in terrestrial carbon stocks associated with land transformations, and
- Emissions from fossil fuel use.

Land Use Changes and Terrestrial Carbon

Land transformations have characterized the entire 10,000-year history of agricultural development and continue on a large scale today. Although concern over deforestation now focuses on tropical areas, many temperate forests have also been cleared at least once during the last few hundred years. CO_2 is emitted in this process, and also when grasslands and savannas are burned to enhance grazing conditions and when carbon contained in soil organic matter is carried off by erosion or converted into CO_2 by microorganisms. Urbanization claims additional forest and agricultural land each year. These land transformations greatly affect how carbon is distributed in organisms, soils, and sediments, and how it flows to and from the atmosphere. The net result has increased atmospheric carbon concentrations, mostly as CO_2 but also in the form of other carbon compounds such as methane.

Today, up to one-fifth of net global CO_2 emissions may be attributed to clearing and burning tropical forests for food production. Additional CO_2 emissions result from burning savannas and agricultural wastes, and using biomass fuels for cooking. Therefore, these activities might have accounted for roughly 8 to 10 per cent of the global warming in the 1980s. As population and economic pressures increase, the rate of deforestation could accelerate. CO_2 emissions from soil erosion may account for about 1 to 2 per cent of global emissions, but data on this pathway are sparse and very uncertain.

Fossil Fuel Combustion-

Most food-related CO_2 emissions in the U.S. occur in the post-harvest phase. However, fossil fuels are also used for food production, for example to drive farm machinery such as tractors and irrigation pumps and to produce, offsite, inputs such as fertilizers and pesticides. Collectively, these uses account for a relatively small share of world fossil fuel use and for about 2 per cent of U.S. CO_2 emissions. *Onsite Fossil Fuel Use:* Global data are relatively poor but suggest that farms released perhaps 76 million metric tons of carbon annually during the mid-1980s through onsite fossil fuel use (not including emissions from fertilizer and pesticide manufacture). This represents about 1 per cent of global carbon emissions from fossil fuel use and accounted for only about one-half per cent of global warming in the 1980s. Similarly, CO_2 emissions from fossil fuel use on U.S. farms during this period (about 14 million metric tons per year) represented about 1 per cent of U.S. CO_2 emissions from fossil fuels. However, both total and per-hectare energy use in the United States have declined sharply since the mid-1970s.

Offsite Nitrogen Fertilizer Manufacturing: Another 1 per cent of world CO_2 emissions from fossil fuel use, or about 0.5 per cent of the global warming in the 1980s, results from commercial fertilizer manufacture. Nitrogen fertilizers account for most of these emissions because they are produced in large amounts and their manufacture (often with natural gas as a feedstock) is very energy intensive. In the United States, nitrogen fertilizer production accounted for about 0.8 per cent of U.S. CO_2 emissions from fossil fuel use. World fertilizer production increased at a rate of 6.2 per cent annually from 1965 to 1985, and the outlook is for continued growth, especially in developing countries.

Post-Harvest Activities

Once harvested, some food is consumed directly by livestock and humans. The bulk, however, is processed, preserved (often though cooking), stored, and distributed with a certain amount of loss and waste during transport. The major emissions from these post-harvest activities include:

- CFCs from refrigeration;
- CO_2 from fossil fuels used in food processing, refrigeration, transport, and cooking, and from biomass fuels used for cooking; and
- CH_4 from decomposition of food-related wastes (including packaging).

Some of these post-harvest activities (e.g., food transport, refrigeration) fall within sectors examined elsewhere in the report but are highlighted here.

CHLOROFLUOROCARBONS (CFCS)

CFCs used in refrigeration are emitted through refrigerant leaks, intentional venting (during manufacture, disposal, or repair), and deterioration of insulation. Worldwide, about 47,000 metric tons of CFC-11 and 85,000 metric tons of CFC-12—roughly 15 per cent of total consumption—are used for food-related refrigeration.

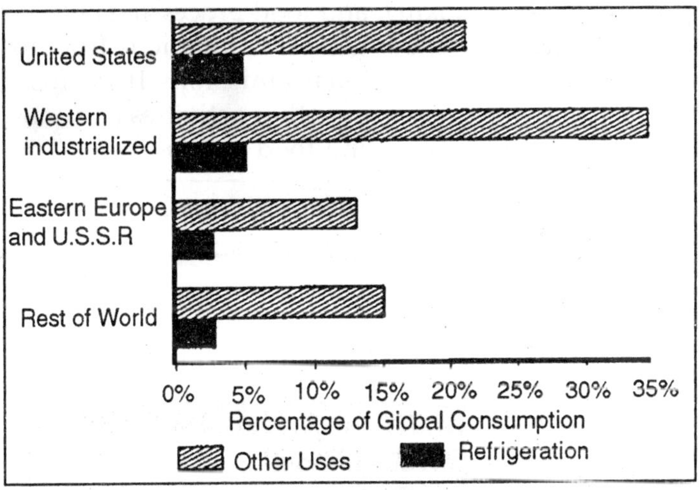

Fig. Consumption of CFC-11 Plus CFC-12 for Food-Related Redrigeration and othe Uses

CFCs used in the food system accounted for about 2 per cent of the global warming in the 1980s, assuming that all are eventually emitted to the atmosphere and that *total* CFC-11 and -12 uses accounted for about 14 per cent (143). Most CFC use in refrigeration occurs in cold storage warehouses, retail refrigeration, and refrigerated transport. Lesser, but still substantial, amounts are used in residential refrigerators and freezers. Global use of CFCs has increased dramatically during the 60 years they have been commercially produced. From 1976 to 1986, despite mounting evidence of the ozone-depleting properties of CFCs, CFC-12 refrigerant sales increased by over 50 per cent. The United States used 41,000 metric tons of CFC-11 and CFC-12 for food refrigeration in 1985. This represents one-third of world use for food-related refrigeration and about 19 per cent of total U.S. use of the compounds.

CFC use is expected to rise rapidly in developing countries. With the promise of funding from the industrialized world, however, key countries such as China and India are expected to join the Montreal Protocol and pledge to reduce and eventually phase out CFC use.

CARBON DIOXIDE (CO_2)

In developing countries, residential cooking is by far the most significant source of post-harvest CO_2 emissions. In industrialized countries, the most significant post-harvest activities are processing in the food industry, residential cooking, and refrigeration (i.e., powerplant emissions attributed to the energy needs of refrigeration). In the United States, energy used for residential and supermarket refrigeration, residential cooking, and food processing and packaging accounted for about 5 per cent of U.S. CO_2 emissions from fossil fuel use, or roughly 1 per cent of global emissions from fossil fuels. Additional CO_2 emissions arise from other activities (e.g., energy use in food wholesaling, restaurants, and food transport).

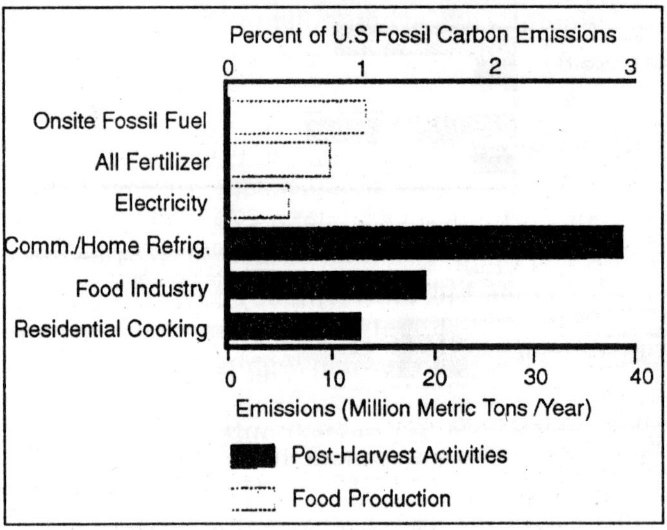

Fig. CO_2 Emissions From Selected Fossil Fuel Uses

In general, more fossil fuel is used in industrialized countries for post-harvest activities than during food production. Even so, CO_2 emissions from post-harvest activities are relatively small compared with those from the energy, building, transportation, and manufacturing sectors.

Food Refrigeration

Accurate estimates of worldwide CO_2 emissions from refrigeration are not readily available, although estimates for specific countries suggest they are important within the food sector but small relative to other sectors.

In the United States, energy use for household and supermarket refrigeration accounted for about 3 per cent of total U.S. CO_2 emissions in 1985. Of this, two-thirds (24 million metric tons) can be attributed to domestic refrigeration, and about one-third to refrigeration at supermarkets. As refrigerators have "saturated" the market in industrialized countries and average energy efficiencies have improved, growth in energy use for refrigeration in these countries has slowed.

By contrast, refrigeration in China accounts for only a small fraction of national energy use (only 0.4 per cent of which goes to generate all electricity), but this is rapidly changing. Between 1979 and 1987, for example, China's production of refrigerators increased by a factor of 125, and many of the several hundred million households that still do not have one may acquire one in the next decade.

Processing, Transportation, and Cooking

In the United States, post-harvest activities accounted for over 19 million metric tons of emissions in 1985, or about 1.5 per cent of U.S. fossil fuel CO_2 emissions; over 40 per cent of this came from purchased electric power. Cooking in residences contributed approximately 12 million metric tons of emissions, roughly 1 per cent of total U.S. carbon emissions. Emissions also result from transporting food, but they are poorly quantified. The overall magnitude of emissions from other sources—such as cooking in commercial establishments and heating of hot water for dishwashing—cannot be readily calculated. In the United States, these emissions could be important given that about half of all meals are prepared outside the home.

In developing countries, cooking is the major source of CO_2 emissions from post-harvest activities and accounts for most household energy use. Energy use in transportation and processing is relatively small. Although CO_2 emissions from cooking have not been quantified, the most common cooking fuels (e.g., biomass, coal) have high carbon content. Traditional biomass fuels (animal dung, crop residues, wood, etc.), which may account for as much as 15 per cent of world energy use, are used extensively for cooking and food processing (e.g., for drying). Coal also is very important in some regions. Over one-fourth of coal use in China, the world's largest coal consumer, is for domestic purposes, primarily for cooking.

METHANE (CH_4)

Following preparation and consumption of food, solid wastes (e.g., food residues, packaging) and sewage are generated. Under some disposal conditions, these wastes result in the emission of CH_4 to the atmosphere. In landfills, for example, carbon-containing compounds decompose in two stages—first aerobically, producing CO_2 emissions; then, once oxygen is used up, anaerobically (i.e., without oxygen) by methane-producing bacteria. Landfills may emit about 30 to 70 million metric tons of CH_4 annually worldwide, about one-half of what livestock emit. In the United States, about 6,000 municipal solid waste landfills were operating in 1986.

ALTERNATIVE PRACTICES

This section discusses alternative practices that could be pursued to reduce future greenhouse gas emissions from the food sector. While the overall costs and benefits of these practices are not clearly defined, collectively they could substantially reduce some emissions. Many would carry other environmental benefits as well—e.g., reducing water pollution from croplands, reducing soil erosion, conserving water supplies, preserving biological diversity, and reducing waste generation in food processing. However, tradeoffs may be associated with some alternatives. For example, some tillage practices used to conserve soils require more pesticides. Also, to reduce the pressure to open new lands to agriculture, crop yields must be increased on existing acreage, which may require greater use of fertilizer and other inputs.

Livestock

Livestock directly produce about 10 to 20 per cent of the world's CH_4 emissions through digestive processes, and indirectly produce additional emissions from anaerobic decomposition of manure. They also indirectly account for emissions of CO_2 and N_2O by virtue of the land and agricultural inputs required to sustain them.

Decrease Methane Emissions Per Unit of Output

Opportunities exist for reducing, or at least limiting, the growth rate of CH_4 emissions from livestock by increasing digestion efficiency and/or animal productivity (i.e., the amount of animal product produced per unit of feed). Emissions reductions on the order of 25 to 75 per cent per unit of product are thought to be possible, with most potential for change in developing countries (industrialized countries have already made strides in raising more productive animals). This range of possible emission reductions roughly corresponds to a 2- to 5-per cent reduction in global CH_4 emissions.

Specific options include:

- Supplement the diets of grazing animals to correct nutrient deficiencies often found in lower quality forage;
- Substitute feeds with low CH_4-producing potential for feeds with higher CH_4-producing potential in the diets of animals in confined (as opposed to free-ranging livestock):
- Develop feed additives that increase digestion efficiency and reduce methanogenesis in the rumen;
- Use growth promotants (e.g., bovine somatotropin); and
- Improve reproductive efficiency.

Enhancing animal productivity can reduce CH_4 emissions given the following assumptions: that by increasing productivity, the same amount of output could be obtained from a smaller herd size; that the rate of output (e.g., milk, beef) per animal increases faster than emissions of CH_4 per animal; and that consumer demand for these products remains relatively stable. It is also assumed that more productive cattle could be brought to market sooner, decreasing the total lifetime CH_4 emissions per animal. the assumption that increased productivity could lead to fewer livestock (and, thus, lower total CH_4 emissions) could be challenged, through. In fact, a large unmet demand for cattle products in developing countries could lead to greater livestock populations even if productivity increases.

Many productivity-enhancing/methane-reducing strategies have been used with great success in the United States and could be transferred to other countries. However, for some countries, especially in the developing world where most cattle graze in unconfined situations, it is difficult to determine how effective some of these options would be.

For example, feed additives called "ionophores," which can increase digestion efficiency, currently are feasible only for confined cattle and therefore may not be of immediate utility in developing countries. Growth hormones, such as bovine somatotropin, have been used in nondairy cattle to increase productivity per animal.

Another option, increasing the reproductive efficiency of animals, could help reduce CH_4 emissions by decreasing the number of cattle needed to produce calves. This could be accomplished by increasing nutrient-use efficiency, as described above, as well as by improving breeding techniques. However, some highly productive breeds developed in industrialized countries may not adapt well to the different environmental and feeding conditions of some developing countries.

REDUCE METHANE FROM ANIMAL WASTES

Manure storage piles, pits, and lagoons are commonly used to reduce runoff from feedlots into surface water and groundwater. Methane could be collected from these sources for later use as a fuel, with technologies such as specially designed biogas generators. About 50 to 90 per cent of CH_4 generated from waste lagoons could potentially be recovered, achieving reductions of up to 1 per cent of total CH_4 emissions. This option is applicable where animals are kept in confined situations, that is, primarily in industrialized countries.

In many developing countries, dried animal manure is an important fuel for cooking and heating; that not used for cooking generally remains in unconfined, pasture/forage systems. However, if manure were collected and processed in anaerobic digesters the CH_4 generated from this process could offset some of the demand for wood fuels.

REDUCE DEMAND FOR LIVESTOCK PRODUCTS

Finally, CH_4 emissions might be reduced by shifting meat production and consumption from ruminants to non-ruminant animals, such as fish, hogs, or broiler chickens, or to more vegetarian diets. For example, to produce a given amount of protein, feedlot beef require nearly five times as much feed as catfish raised in intensive aquaculture systems. Lowering livestock numbers could also help reduce: nitrogenous fertilizer used in growing feed for livestock (and associated N_2O emissions); pressures to expand agricultural acreage in some countries; declines in soil productivity from overgrazing; water pollution from erosion and from runoff of wastes; and health costs of high cholesterol diets, depending on what other foods are substituted in diets.

However, because animal products are also good sources of calcium, iron, zinc, and high-quality protein, reducing their consumption in developing countries may put further nutritional stress on people in some of these areas. In these countries, increased demand for vegetable protein substitutes may also create additional burdens on already stressed grain supplies and have adverse environmental impacts. For example, increased demand for poultry and hogs would require increased amounts of feed and expanded manure handling. Moreover, reducing livestock numbers in developing countries may be especially difficult because of the multifaceted economic and cultural role they play. In many developing countries where livestock are used primarily for draft power and are kept as religious or status symbols, opportunities for emissions reductions may be limited.

Rice Cultivation

In the near term, our ability to reduce CH_4 emissions from rice appears very limited. However, according to a panel of experts convened by the IPCC, a long-term research effort begun today and focusing on new irrigation techniques, more efficient fertilizer use, and developing new high-yield rice species may someday (e.g., two decades) enable global emissions reductions on the order of 10 to 30 per cent.

High-yield varieties grow more quickly, permitting more than one crop to be grown per year. In theory, increasing rice yields per hectare might reduce the need to expand production onto uncultivated lands, thus reducing total CH_4 emissions. However, annual CH_4 emissions per acre will also increase with double harvesting. It is unclear whether increasing harvests on existing lands would result in higher or lower net annual greenhouse gas emissions than clearing new lands for production.

Average yields have not increased significantly since high-yield rice varieties were introduced in the mid-1960s, however, and are not expected to increase dramatically without new technological breakthroughs, for example, genetic engineering to enhance resistance to viruses. Although existing high-yield varieties produce less methane than traditional varieties (because of a higher grain-to-straw ratio), no new technologies to reduce per unit CH_4 emissions are anticipated in the near term.

Better flood control might help increase the production efficiency of high-yield varieties, which tend to show lower and more variable yields under flooded conditions than under controlled irrigation. Fertilizer losses from intermittent flooding would also decline. More research is needed to develop varieties that consistently produce high yields under different conditions (e.g., dry upland environments, rain-fed conditions); and to eliminate the need to flood rice fields where flooding is not natural (e.g., California, South American savannas). Although rice grown under dry-land conditions emits much less methane than rain-fed or flooded rice paddies, dry-land rice accumulates more soil cadmium (a potentially toxic trace metal) than paddy rice. This is a problem in some major rice-producing countries where soil cadmium levels are already high.

NITROGENOUS FERTILIZER USE

If current trends continue, world fertilizer use will double over the next few decades, rising 1.3 per cent per year in industrialized countries and 4 per cent per year in developing countries. Fertilizers often are used very inefficiently; in parts of Asia, for example, fertilizer losses are estimated to be about 50 to 60 per cent of the amount applied. Inefficient fertilization practices result in losses of soil nitrogen through several pathways.

Denitrification is the predominant mechanism of loss, through N_2O formation. The level of N_2O emissions from fertilized soils depends on many factors: fertilizer type and amount, application technique and timing, tillage and irrigation practices, use of pesticides, soil and crop type, and amount of residual nitrogen in the soil. N_2O emissions rates per hectare of cropland can vary by three orders of magnitude depending on how the above factors interact.

Several options are available to increase fertilizer efficiency or reduce the need for fertilizers and thereby reduce N_2O emissions, although the extent to which emissions can be reduced is not known. Nonetheless, other environmental benefits such as reduced nitrate contamination of groundwater and surface waters can also be achieved. These options include:

- Efficient fertilizer application,
- Low N_2O-emitting fertilizers,
- Slow-release fertilizers,
- Nitrification inhibitors, and
- Leguminous sources of nitrogen.

MORE EFFICIENT FERTILIZER APPLICATION

The efficiency of fertilizer use can be increased by determining: how much nitrogen is already available in the root zone as well as how much crops can optimally use; when during the growing cycle fertilizers are most needed; and at what depth they should be placed for various tillage systems. Under certain conditions, for example, fertilizers applied in the spring emit less N20 than those applied during the fall. Efficiency also can be doubled under some conditions by placing fertilizers deep in the soil, rather than "broadcasting" them on the surface.

LOW N_2O-EMITTING FERTILIZERS

The N_2O emission potential of various fertilizers has been studied only under highly site-specific conditions, limiting generalizations about emissions reduction potential of particular fertilizers.

Research is needed on emission levels under a variety of field and cropping conditions. Studies suggest, however, that emission rates may vary by one or two orders of magnitude, with generally higher emissions for anhydrous ammonia than for other nitrogenous fertilizers.

SLOW-RELEASE FERTILIZERS AND NITRIFICATION INHIBITORS

Greenhouse studies with flood-irrigated rice suggest that slow-release fertilizers can considerably reduce denitrification and allow for more efficient plant uptake. Under certain conditions this can double fertilizer efficiency; whether N_2O emissions are simultaneously reduced is unclear.

Also, slow-release fertilizers may continue releasing nitrogen after plants have been harvested, thereby creating the potential for nitrate production and leaching as well as additional N_2O emissions during the winter and early spring. Slow-release fertilizers are not likely to become a viable technology until production costs drop.

Chemical additives in fertilizers can limit soil nitrification processes and, in turn, reduce the amount of nitrate available for denitrification.

A wide range of chemicals has been registered and sold in the United States for use as nitrification inhibitors, and under certain conditions these can reduce nitrogen losses and increase fertilizer efficiency by 30 per cent.

Like slow-release fertilizers, these compounds may only delay the emissions of N_2O. After the plants have been taken out of the ground at harvest and nutrient uptake ceases, more soil nitrogen becomes available for nitrification, which then can lead to further emissions.

LEGUMINOUS SOURCES OF NITROGEN

Nitrogen can be added to the soil by growing "nitrogen-fixing" legume crops such as peas or beans in rotation with grains. A few studies suggest that N_2O emission rates from legume-based systems are similar to those from fertilized crop systems, and possibly higher if no-till practices are used.

If increased use of legumes reduces demand for nitrogenous fertilizers, then CO_2 emissions from fertilizer manufacturing might be lowered. However, the lack of data on N_2O emissions from legume cultivation and on the degree to which legumes could offset fertiiizer use makes it difficult to determine the net effect on emissions. Regardless of their effect on N_2O emissions, the planting of legumes makes sense from a soil conservation standpoint.

Land Use Changes

As mentioned earlier, the food system's single largest contributor to global warming is deforestation. In developing countries as a whole, deforestation is the dominant source of CO_2 emissions. In this section we discuss other ways to encourage land use practices that store more carbon, techniques to maintain or increase yields on existing agricultural lands, and ways to cut production-related food losses and wastes.

Encouraging Transformations That Increase Carbon Storage

In developing countries a great deal of attention has been given to agroforestry—growing trees along with annual crops and livestock. The trees sequester carbon and generate products and revenues for small-scale farmers.

Replacing annual crops on existing agricultural lands with perennial tree crops or woody plants could provide a long-term "sink" for atmospheric carbon as well as produce desirable food products. Examples are hazelnuts and chestnuts in temperate regions and palms in tropical regions. Whereas most of the CO_2 taken from the atmosphere during the growth cycle of annual crops is released again during post-harvest tillage, the roots of woody perennial plants reach much cheaper and lock carbon out of the atmosphere for much longer periods. However, woody crops still require fertilizer and pesticides. Also, new varieties will have to be developed for various cropping systems, and economic and cultural obstacles (e g., development of sufficient market demand, convincing farmers to switch farming practices) must be overcome.

With further research and development, several wild, non-tree perennials, such as eastern gama grass, giant wild rye, and Illinois bundleflower, may provide the germ for future perennial agricultural grains. Like their woody counterparts, perennial grains would conserve soil and water resources. However, the development of perennial grain crops into widely used agricultural staples still may be decades away.

Finally, taking highly erodible agricultural lands out of production and converting them (or allowing them to revert) to perennial grasslands or forests helps conserve soil organic matter, a major carbon reservoir. This practice also helps protect surface waters from agricultural runoff. N_2O and CO_2 emissions may also be reduced through avoided fertilizer and fossil fuel use. Setasides can also increase or help maintain biological diversity.

MAINTAINING OR INCREASING YIELDS

The rate of food production depends on crop acreage and crop yields which, in turn, are determined by a complex set of variables, ranging from soil and plant characteristics to pest outbreaks and varying weather conditions. If per-acre yields are limited or decline, food production can be maintained or increased only by expanding the area of land exploited. In tropical forest areas, for example, peasant farmers commonly respond to declining yields by converting additional forest areas into temporary croplands or by recultivating formerly abandoned areas that have regrown a "secondary" forest. Such transformations are likely to continue unless efforts are made to redistribute land, slow population growth, and stabilize or even increase yields.

Several techniques can be pursued in both industrialized and developing countries to maintain yields while limiting greenhouse gas emissions and limiting area of land used. More efficient fertilization practices and techniques to maintain or increase yields for rice were discussed above. It is important to note that the push to increase yields may require additional fertilizer inputs. As maximum yields are reached, nitrogen efficiencies begin to drop, which could lead to greater N_2O emissions and other adverse environmental impacts.

REDUCING FOOD LOSSES

Food losses from pest damage may cut world food production by one-third and rice production by up to 50 per cent. Adverse weather conditions account for the largest annual variations in food production, which is ominous in view of possible future climate change. Yields also can be unintentionally reduced by human activities.

Efforts could be increased to reduce various wastes and losses in the food system. Techniques for reducing erosion (e.g., conservation tillage, streamside tree plantings) would help maintain productivity. Post-harvest losses and wastes from pests, spoilage, and other factors could be reduced in several ways. Nutrients from human wastes (e.g., food residues, sewage) can be recycled back into the food production system, rather than relegated to landfills or discharged into surface waters. Treated wastewater from sewage treatment plants is now being used for aquaculture and for irrigation water in countries like Israel.

Recycling food wastes and sewage wastewater can help improve soil quality, thereby reducing the need for supplemental fertilizer and other energy inputs. However, the costs of processing and transporting wastes and the problems associated with chemical and biological contaminants in the wastes pose disadvantages.

CFCs, CO_2, and Refrigeration

New refrigeration systems are significantly more energy efficient than older systems and, hence, emit less CO_2 from electricity requirements. However, efficiency improvements have not led to comparable reductions in CFC emissions; indeed, these improvements are partially attributable to greater use of CFCs in insulation. Political pressure, however, is building to reduce both CO_2 and CFC emissions, and some new systems and components can reduce or eliminate CFC use (both as a refrigerant and as insulation) and reduce CO_2 emissions. Some promising systems involve highly effective CFC-free insulation that improves energy efficiency. Using smaller refrigerators also can reduce CFC and CO_2 emissions.

Other opportunities include using different working fluids and energy sources. For example, refrigerators can operate on energy sources such as natural gas, solar energy, and heat generated from waste materials, all of which would reduce CO_2 emissions and energy costs. The major drawback is the capital cost of shifting to new technologies.

Emissions from CFCs already in use as refrigerants could be limited by minimizing accidental emissions during repair (e.g., from leaks) and by sequestering and/or destroying CFCs instead of venting them during repairs or prior to final disposal of refrigeration systems. The primary drawbacks are the costs of recovery and disposal and the costs of purifying CFCs for reuse.

Existing inefficient equipment also could be retired early in order to accelerate deployment of better technologies. The advantages include the possibility of rapid implementation, reduced fuel costs for new equipment, and, in some cases, reduced costs for electric utilities, which may be spared building additional generating capacity. The primary disadvantages are the costs of collection and disposal and of purchasing a replacement unit.

Cooking, Food Processing, and Packaging

Emissions from cooking can be reduced, in theory, by changing the types of energy used or by improving fuel-use efficiency. Any strategy that promotes these changes, though, must consider fuel costs and availability, how cooking is normally conducted, and dietary and social preferences. In addition, improved cooking efficiencies may simply allow more cooking with the same amount of fuel, with no substantive change in emissions. Because these considerations have not been fully explored in different parts of the world, assessments of the global potential for limiting emissions from cooking cannot be made readily.

The United States and other industrialized countries emit far less CO_2 from cooking than from other activities. Even so, emissions can be reduced by using more efficient ovens and stoves. In developing countries significant CO_2 and other emissions result from cooking with coal and biomass. Shifts to other fuels have occurred fairly quickly in some cases. Increases in cooking efficiency also have occurred—often through improved cookstove designs—and can significantly reduce emissions per unit of delivered energy; however, they do not necessarily reduce total emissions. Where fuel availability is already constrained, improved efficiencies may allow people to cook more or to shift fuels to other end-uses such as heating.

There are also opportunities in the areas of food processing and packaging to improve energy efficiencies and switch to low-emission energy sources. Options range from gas-fired cogeneration of electricity and process heat to more fundamental process modifications.

CO_2 and Machinery, Fertilizer Manufacture and Irrigation

Although the impact of reducing emissions from farm machinery (e.g., engines, pumps) and fertilizer manufacture would be relatively small, technologies to improve energy efficiency could help reduce reliance on fossil fuels and hence save farmers considerable expense. Promising options to reduce fossil fuel use during food production involve changes in the character and efficiency of field operations (e.g., more efficient farm vehicles and irrigation, use of ethanol fuels, and innovative tillage practices and crop drying techniques) and improved efficiency in fertilizer manufacturing. Alternative energy sources such as wind and solar could be used for some operations).

EFFICIENCY IMPROVEMENTS FOR FARM VEHICLES

Over the next 10 to 15 years, for example, farm-tractor fuel efficiencies could be improved by 5 to 15 per cent with new technologies such as adiabatic engines equipped with turbochargers, electronic controls, and onboard system diagnostics. However, these technologies require upfront capital expenditures and their long-term reliability is unknown. In developing countries their potential impact is unclear. In addition to costs, some analysts suggest that mechanization will be slow because tremendous labour supplies exist.

FERTILIZER MANUFACTURE

Energy efficiency in fertilizer manufacturing has improved substantially. By the mid-1980s, new plants were using about 20 per cent less energy than in the early 1980s primarily due to energy recovery equipment. Several new urea processes could decrease energy requirements by another 25 to 50 per cent relative to the U.S. plant average. CO_2 emissions reductions gained by improving energy efficiency could be negated, however, if more coal replaces gas as a primary feedstock in the production process.

EFFICIENT IRRIGATION

The energy intensity of irrigation in the United States continues to rise, primarily because of increased pumping of groundwater; the same is true for some developing countries such as India and China. For U.S. food production activities, energy use for irrigation ranks third (behind pesticide and fertilizer manufacturing and use, and farm machinery use). Worldwide, more than 60 per cent of irrigation water, on average, is lost due to inefficient practices. Technologies available to reduce water and energy use in irrigation include: Low Energy Precision Application (LEPA) designed to apply irrigation water and agrichemicals in small amounts and in precise locations; sprinkler and drip-irrigation systems to reduce evaporation; monitoring of soil moisture so water can be applied when needed; liners in canals to prevent seepage; and lasers to measure field levels so that water can be evenly distributed. Recycling of agricultural runoff and municipal wastewater can also reduce demands for irrigation water, but energy requirements for pumping may be high.

ETHANOL FUELS

As discussed in the transportation sector, corn-based ethanol emits from 10 per cent less to 30 per cent more CO_2 than gasoline. In 1987, about 3.2 billion liters of ethanol were sold in the United States, making this country the world's second largest ethanol consumer after Brazil. Over 80 per cent of U.S. ethanol plant capacity in 1986 was dedicated to corn feedstocks. Although the above estimates do account for the additional CO_2 emissions associated with the manufacture of fertilizers, pesticides, and other energy-intensive inputs needed for increased corn production, other impacts—such as additional N_2O emissions from fertilizer breakdown, increased soil erosion, and other environmental problems associated with corn crops grown in monoculture must also be recognized.

INNOVATIVE TILLAGE PRACTICES

By simultaneously laying seed and herbicides onto unplowed soil, a farmer can limit tractor trips to just one per crop cycle. This can reduce fuel use and attendant emissions, as well as enhance the soil's ability to retain organically bound carbon and water. This and other "conservation tillage" practices are primarily used to control soil erosion. However, they tend to require more herbicides for weed control than conventional tillage and therefore may result in greater N_2O emissions. They also require more seed.

CROP DRYING

In the United States, most crops sun-dry in the field. Some crops are dried with heated air, though; this accounted for about 3 per cent of total on-farm energy use in 1978, mostly for corn and tobacco. Liquefied petroleum gas and natural gas are the most common energy sources. Alternative sources such as solar energy can reduce fuel use by as much as 20 per cent. At present, 2 per cent of U.S. crops are dried using active solar energy systems. Many passive solar crop drying systems (i.e., systems that rely on natural air convection) are in use in developing countries.

POLICY ISSUES AND OPTIONS

Most options discussed in the preceding section individually provide relatively small potential for reducing greenhouse gas emissions. But the sum of such efforts may someday provide substantial emissions reductions. In the case of livestock and CFCs for refrigeration—categories that together accounted for about 6 to 9 per cent of the global warming in the 1980s—promising, substantial opportunities may exist in the near term. While reductions in CH_4 emissions from rice cultivation are theoretically possible, technologies to achieve this are much farther off and will require significant research and development. Techniques to increase fertilizer-use efficiency are currently available and will help reduce localized water pollution problems, but their effect on N_2O emissions is inconclusive and will require more study. Improvements in fuel efficiencies of farm equipment and fertilizer manufacturing are likely to result in minor CO_2 emissions reductions in developed countries, even less in developing countries where the level of mechanization is low and is likely to remain so in the near future.

This discussion focuses on policy mechanisms that Congress could use both to implement some of the technical options discussed above for the U.S. food system and to influence international emission reduction efforts.

Policy considerations extend beyond technological factors to include many social, political, and economic issues. In the United States, for example, choices will be greatly influenced by farm support programs. In many developing countries, efforts to limit food sector emissions and to gain associated environmental benefits will have to be linked with efforts to combat poverty, inadequate nutritional levels, inequitable land distribution, and rapid population growth.

Research Issues

One of the most important research priorities for understanding the relationship between the food system and global climate change is the development of an emissions database representative of agricultural systems and growing conditions throughout the world. For example,

- Better CH_4 emissions data are needed from the large rice-producing areas in Asia and the Pacific; the general biogeochemistry of CH_4 production in flooded rice paddies also needs to be established;
- Factors affecting CH_4 emissions from rice cultivation, such as climate, soil and water, species type, use of fertilizers, cultural practices, site, seasonal and diurnal variations, and relationship to other greenhouse gas emissions (e.g., N_2O), need to be studied to establish representative emission factors;
- The relationships between N_2O emissions and natural factors; fertilizer type, application rate, and placement; residual soil nitrogen; crop-specific nitrogen uptake; soil and water conditions; and timing need to be established;
- Uniform, simple, and inexpensive techniques for measuring CH_4 emissions from rice paddies and N_2O emissions from all types of fertilized soils must be established so that comparable data can be collected worldwide; and
- How CH_4 emissions from livestock vary by type and age of the animal and by type of management system (e.g., how and what animals are fed; manure handling) needs to be enumerated for the many regions throughout the world with large livestock populations.
- The relationship between biomass burning and emissions of trace gases (including CH_4, N_2O, NO, and others) and the

effects of such burning on the atmospheric and terrestrial environments.

Many international organizations already fund or coordinate agricultural research (e.g., International Fertilizer Development Centre, International Board for Soil Research and Management, International Council for Research in Agroforestry, Inter-American Institute for Cooperation on Agriculture). However, there is no overall promotion or coordination of research on the relationship between agriculture and global climate change. The United States could promote an international programme to focus greater attention on this issue and to develop research protocols so that results can be meaningfully compared on a global scale. The Consultative Group for International Agricultural Research (CGIAR), an association of 13 regional and international agricultural research centers, might appropriately house such an effort.

Livestock

U.S. Practices

On the domestic front, Congress could direct the U.S. Department of Agriculture (USDA) to determine the extent to which methane-reducing techniques, such as feed additives, ionophores, and other nutrient management techniques, as well as animal waste management, are currently used in the United States. Such a programme could also identify both institutional and technical barriers that hinder more widespread development and use of such techniques. Congress could provide additional support (e.g., through the USDA Agricultural Research Service and the National Science Foundation) for research on these techniques.

To limit future growth in, or even reduce, livestock populations in the United States, Congress could reduce or remove price supports for feed grains, which might make beef and dairy products more expensive (although it is unclear if the costs would rise or fall in the long term). About two-thirds of the total Federal grain subsidies apply to livestock feed. Feed grain farmers might grow other crops that make more money and, if feed grain prices rose sufficiently, livestock producers might raise less meat. However, this could cause large near-term economic disruptions for some farmers and portions of the food industry.

Congress could also modify eligibility criteria in the Conservation Reserve Programme so that farmers can choose to put more land now used to grow feed grains into reserve; this also would reduce CO_2

emissions from onfarm fossil fuel use and fertilizer manufacture and N_2O emissions from fertilizer use.

Developing-Country Practices

In developing countries, programs to increase productivity through improved breeding techniques or to enhance animal waste management systems must meet the special needs of livestock management systems in these countries, where livestock are primarily pasture-fed and are used for many purposes other than provision of food. With this in mind, Congress could contribute funds, through U.S. bilateral aid programs and through multilateral organizations, to expand research programs in developing countries so that methane reductions become an additional research priority. For example, research institutes such as the International Livestock Centre for Africa and the International Laboratory for Research on Animal Diseases are part of the CGIAR system, which receives U.S. funding through A.I.D.

However, promoting technologies that lower per-animal emissions or policies that reduce livestock numbers will probably be difficult. Many of the technologies designed to reduce methane from individual animals are geared toward controlling the diets of animals in feedlot management systems, which are less common in developing countries. Also, efforts to introduce more productive livestock breeds into developing countries must first recognize that the unique genetic qualities of indigenous breeds that have evolved over thousands of years in adaptation to different ecological conditions.

Because livestock in developing countries are used for many purposes other than food production—as symbols of social status and wealth, for their religious values, for draft (construction) activities, for the energy value of their manure, and as alternative sources for income in the event of crop failures—convincing peasants to change their livestock management habits or to reduce their livestock numbers will probably be difficult. Also, it may be difficult to decrease the lure of cattle ranching to middle- and upper-class landowners and investors. In many countries (particularly in Latin America), ranching is encouraged by national development policies, land ownership patterns, and land speculation. Indirect opportunities exist for Congress to influence this particular situation, through its control of funding for bilateral aid programs and influence on multilateral lending institutions, but many obstacles must be overcome.

CFCS, CO_2, AND REFRIGERATION

In industrialized countries, obstacles to implementing energy efficiency options and other refrigeration improvements include consumer attitudes, regulatory barriers, and technical problems (e.g., lack of equipment that can directly use replacements for CFCs). The basic issues, though, are not whether refrigeration can be accomplished more efficiently and without CFCs, but how best to do this and at what costs compared with the benefits of emissions reductions.

Steering developing countries away from CFC production, CFC-based refrigerators, and low-efficiency equipment will be more problematic. Policy alternatives include encouraging these countries to sign and ratify the Montreal protocol, and transferring information, technologies, and capital to enable them to pursue alternative and acceptable refrigeration practices economically.

NITROGENOUS FERTILIZER USE

Congress could promote more efficient fertilizer use in the United States by changing commodity programs so as not to encourage excessive production and to allow farmers to grow crops and adopt practices that rely less on commercial fertilizers and other energy-intensive inputs, without loss of programme base acreage.

Best Management Practices (BMPs) are designed by the Soil Conservation Service (SCS) to reduce soil degradation and water contamination from agricultural activities. At present, the SCS does not have statutory authority to promulgate enforceable regulations. Congress could require implementation of BMPs through cross-compliance, i.e., make implementation a prerequisite for receiving Federal price and income supports. However, such a policy would not apply to the one-third of U.S. croplands that are not enrolled in Federal farm support programs. Congress also could provide incentives (i.e., special services from USDA extension agents) to farmers who voluntarily adopt BMPs.

Carbon Dioxide and Land Use Changes

Encouraging Land Use Changes That Increase Carbon Storage

The goal of the USDA's Conservation Reserve Programme (CRP) is to take 40 per cent (16 to 18 million hectares) of highly erodible croplands (about 10 per cent of all cropland) out of production, and in some cases to plant trees or grasses on the land.

Farmers who take lands out of production for 10 years receive annual rental payments from the Federal Government. By the end of 1989, about 14 million hectares had been enrolled at a cost of over $1 billion annually; the Food, Agriculture, Conservation, and Trade Act of 1990 (Public Law 101-624) extended the sign-up period through 1995.

The USDA estimates that the CRP could reduce U.S. fertilizer use by as much as 3 per cent.

Congress could modify the CRP by:

- Increasing the acreage goal;
- Including croplands eroding at moderate levels;
- Including other environmental objectives such as groundwater protection;
- Increasing incentives for enrollment (e.g., provide options to extend leases, lengthen lease periods, and/or increase rental payments); and
- Providing incentives for managing existing croplands in an environmentally sound manner.

All of these options will require additional programme appropriations. An expanded CRP could also have detrimental economic effects on communities that depend on local farm business (e.g., farm equipment dealers, repair shops, agrichemical dealers, etc.), and consumers would likely protest if food costs went up substantially.

DISCOURAGING LAND USE CHANGES THAT INCREASE EMISSIONS

Within the United States, limiting land transformations is primarily a local or State zoning issue. However, some Federal benefits (e.g., housing and infrastructure grants) and regulatory permits (e.g., for industrial facilities) affect the disposition of agricultural land. Efforts to consider long-term environmental issues in land-use decisions could be important symbolically for international attempts to influence land use decisions in developing countries. Efforts to slow urbanization can also reduce urban infrastructure costs, limit automobile travel, and otherwise contribute to more livable and affordable communities.

MAINTAINING OR INCREASING YIELDS

Congress could require USDA to expand existing programs (i.e., "Low-Input Sustainable Agriculture") and develop new ones that focus on alternative practices, including techniques that maintain or increase crop yields and reduce emissions per unit of food output. Congress could also increase research funding to define relationships among agricultural practices, crop yields, and emissions, and change existing U.S. domestic agricultural commodities programs that discourage farmers from pursuing alternative technologies and methods such as crop rotation and integrated pest management.

Congress also could promote alternative practices overseas, particularly in developing countries, by increasing support for:

- A.I.D. assistance programs in sustainable agriculture (e.g., technical assistance, research and development); and
- Multilateral programs such as those of the Food and Agriculture Organization, CGIAR, and numerous other international agricultural research institutions.

Projects funded through these sources must recognize, however, that alternative agricultural practices developed by, and for, the industrialized world may not be the most appropriate for the developing world. For example, in the humid areas of the developing world, significant amounts of harvested crops are lost due to inadequate storage, while in other countries post-harvest losses may occur for different reasons.

CO_2 AND MACHINERY, FERTILIZER MANUFACTURE, AND IRRIGATION

Federal options for promoting efficiency in the production and use of energy for agriculture are numerous. They include increasing the cost of energy; setting efficiency standards; supporting research, development and demonstration projects; and providing incentives to retire old equipment and deploy low-emission alternatives.

NITROGEN AND CORN PRODUCTION

An adequate supply of plant-available nitrogen (N) is crucial for efficient corn production, and corn N requirements are greater than any other nutrient. For example, a corn crop yielding 150 bushels per acre typically contains about 165 lbs N in the grain and stover, or approximately 1.1 lbs N/bu grain. These calculations are based on actual N uptake, and allowances must be made for actual fertilizer use efficiency and soil N availability.

Nitrogen use efficiency is the percentage of applied N that is actually taken up by the crop. Nitrogen use efficiency is normally 50-60%, but can be as high as 75% with proper N timing and placement. Costs of N fertilizers and environmental concerns about nitrate (NO_3) from fertilizers, manures, and other nutrient sources leaching into groundwater require that new tools be developed and implemented to improve the use of all N sources available to the corn crop.

Many sources of N can be used by a corn crop. Residual N is N that is carried over in the soil from one growing season to the next, and this N source may supply as little as 10% or as much as 100% of the crop's total N need. Residual N is generally low under our climatic conditions because it tends to leach out of the root zone as NO_3-N during the winter and early spring months; however, residual N can meet a significant portion of a subsequent corn crop's need in silty and clayey soils, which retain more N than sandy soils. Most residual N is found in organic forms such as animal manures, legume forage and cover crops, and biosolids (municipal wastewater treatment sewage sludge).

The N that is in the organic forms becomes available to the crop as the organic matter is mineralized (decomposed) by soil organisms. Mineralization of organic matter and release of N increases as soil warms in the spring. When little N is available from animal manure, biosolids or legumes, the majority of the crop N requirement must be supplied by mineral fertilizer sources such as urea ammonium nitrate solution, urea, ammonium nitrate, or ammonium sulfate.

NITROGEN BEHAVIOUR AND SOIL TESTING

Soil testing has been an economically and environmentally beneficial practice for determining the availability of phosphorus, potassium, and other nutrients. Nitrogen recommendations for corn grown in the middle Atlantic and southeastern states have not been based on traditional soil test calibration methods, in contrast to other nutrients, because: 1) predicting the availability to crops of organic N has not been successful, and 2) the storage of soil mineral N is often too brief due to the winter and spring leaching losses of the mobile NO_3-N form that occur under our high rainfall and moderate soil temperatures. Nitrogen analysis of animal manures, biosolids, and legumes to predict the amounts of N available for crops has been only partly successful, which accentuates the need for a soil N test.

Nitrogen behaviour in the soil is difficult to predict because N transformations in soil are very complex. Over 98% of the N in most soils is unavailable for plant uptake at any specific time because it is fixed in soil organic matter or in clay minerals. Nitrogen in organic matter (e.g., plant residue, cover crops, animal manure, biosolids) may undergo microbial transformations that convert it to a plant-available form. The end result of this process is the NO_3 form of N. The transformation rate of organic N to plant-available NO_3 in the spring increases as the soil temperature increases, and much plant-available N can be produced from organic N in Virginia, beginning in mid-May in eastern Virginia to mid-June in western Virginia.

The behaviour of N in the soil has several important implications for efficient N management of corn. Corn requires only small amounts of N during the first month of growth because the plants are small and root systems are not well-developed. Nitrogen applied preplant or released from organic matter during early spring can be lost by leaching during this time when the plants' N requirements are low and soil moisture is high. Therefore, only small amounts of starter N (25-30 lbs/acre) should be applied prior to or at corn planting to meet the N needs of the crop for the first 30-45 days following emergence. Any additionally-required N can be applied as a sidedressing when the corn is 12 to 24 inches tall.

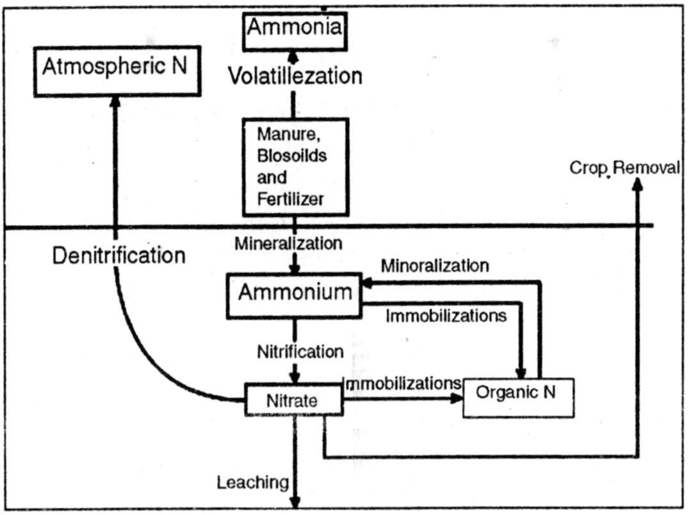

Fig. The Nitrogen Cycle in Soil. (Freelance)

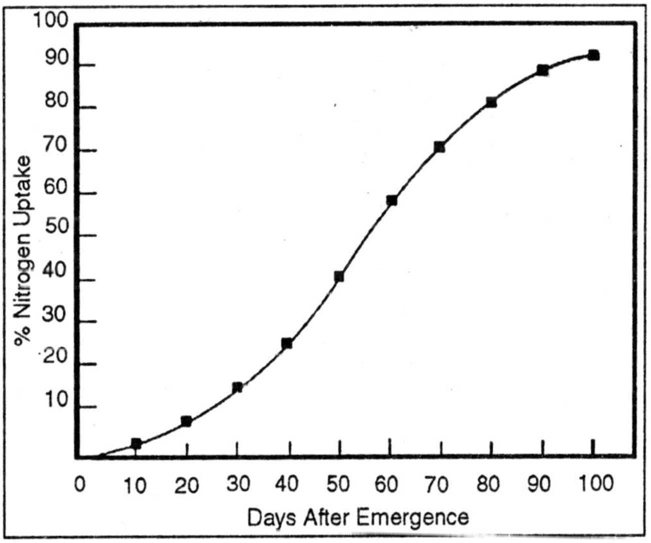

Fig. Nitrogen Uptake by Corn. (Origin)

Soil testing in Virginia for available N before the growing season, as is practiced for other nutrients, does not accurately reflect the availability of N when it is most important to the crop (i.e., 30 to 45 days after emergence). The NO_3-N soil test procedure relies on sampling and testing after the crop has emerged and grown for several weeks. The concentrations of soil NO_3-N measured by this procedure are the result of many complex reactions affecting soil N and are more closely related to the need for supplemental N fertilization than any other procedure tested to-date.

PRE-SIDEDRESS SOIL NITRATE TEST

The pre-sidedress soil nitrate test (PSNT) is based on sampling of the surface one foot of soil after the soil has begun to warm and before the corn begins its most rapid growth rate (i.e., corn is 10 to 15 inches tall at the whorl). The amount of NO_3-N in the soil sample is an accurate index of plant-available N, and sidedress fertilizer N recommendations can be modified depending on the concentration of NO_3-N found in the soil.

Data from 47 field research experiments conducted in the Coastal Plain, Piedmont, and Ridge and Valley soil provinces in Virginia during 1990 and 1991 demonstrated that corn grain yields were maximized (i.e., relative yield = 1.0, or 100% of maximum) at soil NO_3-N concentrations above 18 parts per million (ppm). Enough N was mineralized, or made available from decomposing organic N, at the locations where NO_3-N concentrations were above 18 ppm to supply the seasonal N needs of the corn. Below 18 ppm, most of the relative yields were low, and supplemental N was needed to attain maximum yields. High relative yields occurred primarily where soil had received considerable contributions of organic N, such as from animal manures, biosolids, legume forages, or legume cover crops. Therefore, the PSNT should be utilized primarily on soils that have received significant amounts of organic N.

Index

A

B

C

D

E

F

G

H

I

K

L

M